# Things New & Strange

# Things New & Strange

## A Southerner's Journey through the Smithsonian Collections

G. Wayne Clough

*Foreword by Lester D. Stephens*

The University of Georgia Press *Athens*

Published in association with the Smithsonian Institution

Smithsonian Institution

Athens, Georgia 30602
www.ugapress.org

Designed by Erin Kirk New
Set in Adobe Caslon Pro
Printed and bound by Versa Press
The paper in this book meets the guidelines for permanence and durability of the Committee on Production Guidelines for Book Longevity of the Council on Library Resources.

Most University of Georgia Press titles are available from popular e-book vendors.

Printed in the United States of America
23 22 21 20 C 5 4 3

Library of Congress Cataloging-in-Publication Data

Names: Clough, G. Wayne, author.
Title: Things new and strange : a southerner's journey through the Smithsonian collections / G. Wayne Clough.
Description: Athens [Georgia] : The University of Georgia Press ; [Washington, D.C.] : Published in association with the Smithsonian Institution, [2019] | Includes bibliographical references.
Identifiers: LCCN 2018058786 | ISBN 9780820355238 (hardcover : alk. paper)
Subjects: LCSH: Georgia—Civilization. | Georgia—Social life and customs. | Clough, G. Wayne—Childhood and youth. | Smithsonian Institution.
Classification: LCC F291.8 .C55 2019 | DDC 975.8—dc23
LC record available at https://lccn.loc.gov/2018058786

# Contents

# Foreword

The Smithsonian Institution is one of America's greatest treasures. Comprising nineteen museums and galleries, the National Zoo, and nine research centers and holding 154 million natural history specimens and cultural objects plus thousands of books, documents, and films, this vast organization requires a director of broad vision and extraordinary ability. In 2008, the Smithsonian Board of Regents appointed Dr. G. Wayne Clough to fill that prestigious position. An esteemed scientist and the long-time president of Georgia Institute of Technology, Dr. Clough became the twelfth secretary of the Smithsonian and the first person born in the American South to hold the post. It was the confluence of his service as secretary and his affection for his native South Georgia that brought about this book, an interesting adventure into two different but connected worlds.

Although he left his birthplace at the age of twelve, Dr. Clough maintained a strong attachment to Douglas, Georgia, his original home. Following his retirement from the Smithsonian in 2014, he decided to write about his early years in Douglas and about the broader area of South Georgia. At the same time he planned to look for connections between the region and the institution he had guided for six years. The challenge was how to merge the two interests. He succeeded in the endeavor and produced this insightful exploration, which deftly weaves a seamless story from his life in Douglas, to the southern section of the state, and into the Smithsonian collections of animal, plant, fossil, and rock specimens; cultural artifacts; and works of art. Aiding in his quest were expert curators in each museum section and old friends and other people residing in the vicinity of his childhood home.

In an engaging style, Dr. Clough introduces us to his native area and the influences that shaped him there during the 1940s and early 1950s, and he guides us through various collections in the Smithsonian. It is a pleasure to share his fascinating finds. He identifies ties between the institution and South Georgia, including extinct fauna, geology, indigenous peoples, once-enslaved humans, history, and culture. Adding to the interest of his account are the many Georgians linked to the Smithsonian, among them former president Jimmy Carter, who has an important place in this book.

A blend of memoir, views on regional character, and comments on the nature of a national institution, this book constitutes a delightful intellectual excursion. Dr. Clough's compelling curiosity has resulted in increasing our understanding of South Georgia and the Smithsonian Institution and, as he hoped when he began this journey, in identifying the existence of a sure bond between his home region and the great national institution he once led.

Lester D. Stephens
Emeritus Professor of History
University of Georgia

# Things New & Strange

# Introduction

One of my first official duties as secretary of the Smithsonian Institution was to be blessed by Buddhist monks. The ceremony took place in a wooden replica of a Buddhist temple, built on the National Mall for the Smithsonian's 2008 Folklife Festival, which featured the small Himalayan nation of Bhutan. Representing Bhutan at the festival's opening ceremonies were Prince Jigme Dorji Wangchuck of the royal family and a group of Buddhist monks from a remote mountain monastery. When it was over, the monks told me through an interpreter that they had flown on an airplane for the first time on their trip to Washington, D.C. They said it felt like an out-of-body experience, and as the brand-new Smithsonian secretary, I could relate. But my journey had begun many years before in a small town in South Georgia.

Some may have wondered about my qualifications to be secretary of the Smithsonian Institution, which is the equivalent of being the CEO of a corporation. I could not blame them, because hardly anyone is fully prepared for the diverse responsibilities that come with this position. The Smithsonian is a multidisciplinary enterprise with activities in 140 countries; nineteen museums and galleries plus the National Zoo; and nine research centers, including the Smithsonian Astrophysical Observatory, which operates observatories around the world. The Smithsonian serves 30 million visitors a year and reaches about 150 million people through its television channel, magazines, websites, and social media. Not to be left out are the collections, which contain more than 154 million objects, specimens, and works of art. And this total does not include its archives and libraries, whose books, documents, recordings, videos, and films number in the millions.

You don't apply to become the secretary of the Smithsonian; the job finds you, because someone believes you are the type of person who can preside over an institution that complex and historic. It helps if you like a challenge, feel that you owe a debt to your country, enjoy learning, and are prepared to devote evenings and weekends to the job. Someone believed I was such a person, and the Board of Regents of the Smithsonian agreed. I was the first secretary born below the Mason-Dixon Line.

My home was the small town of Douglas, the seat of Coffee County in southeastern Georgia—far enough south to be closer to Jacksonville, Florida, than to Atlanta, the capital of Georgia. Asked where we were from, my family always said "South Georgia," a place where the land is flat, often sandy, traversed by slow-moving rivers, and dotted with swamps. To us, North Georgia, although part of the same state, was a different place—home to hills and mountains, fast-moving rivers, and the big city of Atlanta. My first trip through North Georgia did not come until I was twelve, when we drove through that part of the state on our move to Chattanooga, Tennessee. I was amazed to see my first mountains and, in the winter, my first snowfall.

When I was growing up in Douglas in the 1940s and 1950s, our schools were segregated; streets such as Grady Avenue, where I lived, were not paved; and most folks who lived on nearby farms, including many of my relatives, had no electricity or indoor plumbing. History books in our schools referred to the Civil War as the War between the States, and a memorial to the Confederate soldier stood at the center of town. It was also a place where a young boy could immerse himself in nature—roaming the pine barrens, exploring the swamps, swimming in the rivers, camping under the stars, and learning about flora and fauna through firsthand encounters.

After we moved from Douglas to Chattanooga, I continued my love affair with nature, only now in the surrounding Appalachian Mountains. Geology fascinated me, and I took up spelunking: exploring the caves that were found in the limestone formations. Spelunking led to my studies in geological engineering, first at the Georgia Institute of Technology and eventually at the University of California, Berkeley, where I earned my PhD. I embarked on an academic career and was fortunate to serve as a faculty member and administrator at a number of prestigious universities. At the culmination of my academic career I returned to my undergraduate alma mater, Georgia Tech, as president. As I contemplated retirement after fourteen years in that post, fate swooped in, and I was appointed secretary of the Smithsonian. Although some people scratched their heads over the amazing fortune of a small-town boy, it was actually the small town in the boy that served as the anchor throughout the unexpected experiences of a peripatetic life.

As my tenure at the Smithsonian drew to a close, I thought about writing a memoir of my early years, but I realized that others had already done the "southern childhood" book better than I ever could. However, as I thought about this idea, it became apparent how little I really knew about the place where I had spent my childhood. That's when it struck me. I was the head of an institution with vast collections that covered natural history, pre-Columbian history, American history, art, culture, and just about everything else. Why not explore the insights the Smithsonian collections could provide on the place where I grew up? Like many apparent strokes of inspiration, this one would prove to be a bit like Alice's rabbit hole—once inside, there was no going back.

The big historical events that shaped our nation barely grazed my hometown of Douglas. So, as I began my quest to learn about my boyhood home from the Smithsonian's collections, I worried that not much from Douglas or South Georgia had merited inclusion. But this concern showed how little I understood about where I grew up and how much the Smithsonian collections had to teach me.

When I began to search the collections, I was pleased to discover more artifacts connected to South Georgia than I ever expected, and most of them had people tied to them. Usually, people entered the equation with the acquisition of an artifact. Sometimes they were Smithsonian staff who were focused on adding to the collections. But just as often they were unusual people who collected out of love, a desire to learn, a hope they could make money, or a little of all of these things. Collectors can be obsessive and may go to great lengths to find, buy, dig up, or even occasionally steal what they want. The portraits, books, landscape paintings, and photographs in the history and art collections also tell interesting stories about people. And once artifacts are in the collections, another group of people—curators—takes care of them, organizes them, writes books about them, or makes revelatory findings with them. The people tied to the collections become part of the stories taught by the collections.

Along the way many asked about my project, and I realized that other people might want to undertake a similar search. After all, here was a means to learn about almost anything from a different point of view and an opportunity to experience the surprise and joy of seeing things in new ways. Of course, as the secretary of the Smithsonian I had access to any curator I wanted to ask for help. But I am convinced that the digitization of the collections, which is well underway and proceeding rapidly, offers everyone the opportunity to pursue their own quests through direct online access to objects and documents. The possibilities are unlimited.

A recent concept in our digital world is the Internet of Things. With the development of the cloud and linked digital systems, everything and just about everybody can somehow be linked to everything and everybody else. I learned that the concept of interconnectedness is true when it comes to the Smithsonian collections, if you dig deeply enough. As the great naturalist John Muir said, "When we try to pick out anything by itself, we find it hitched to everything else in the universe." It was this idea that made me realize that my search through the collections was not about understanding what my home in South Georgia was, but rather about understanding how it came to be what it was and why.

## CHAPTER ONE

# Recollections of a South Georgia Childhood

It was the spring of 2014, and we were traveling from the Savannah airport to Douglas, when the flashing blue lights of a Georgia state trooper suddenly appeared behind us. Johnny Gibbons, the science media specialist at the Smithsonian, had volunteered to drive our rental car, and I had warned him to slow down in the small towns that make a living from speeding tickets. However, we were late for dinner with friends of mine in Douglas, and there had seemed no reason not to take advantage of this deserted stretch of highway.

Johnny is a native of New York state, tall and redheaded, nothing southern about him, so I told him to let me do the talking. The trooper said we were doing seventy-five miles per hour in a fifty-five zone. Putting on my best South Georgia accent, I explained we were late to see some dear friends who were holding dinner for us and were profoundly sorry for speeding—which we were, given the circumstances in which we found ourselves. The trooper looked at Johnny and then me, thought about it, and said, "Okay, sir, I'll let you off with a warning. But you need to call your friend in Douglas and tell him you'll be late, because you have to slow down." Both Johnny and I promised to call, but when we tried, our cell phones showed no service.

The point of this pilgrimage to South Georgia was to see my childhood home with new eyes. Of course, I had occasionally returned as an adult to visit my parents, who had moved back to Douglas after retiring, and to attend the funerals of relatives. But time constraints did not allow for introspection on those occasions. With Mom and Dad no longer living, I was now unbound from my former obligations and would have an opportunity to re-experience memories I had stored away.

By the time I began contemplating this book, many of my childhood memories had settled into the deeper recesses of my mind. What remained vivid years after I left my boyhood home were sensory impressions. I could still feel the hot summer sand on my bare feet, hear the breeze whispering through the needles of the longleaf pines, see the sun glint off the dark waters of a swamp, and smell the tobacco curing in the barns. But the memories that floated to the surface seemed without substance.

It was time to go back and see the places where I grew up, renew my acquaintance with people who never left, and revive memories. A visit would also offer a chance to explore the potential for linkages between South Georgia and the Smithsonian collections. It turned out to be the first of many return visits.

After our encounter with the state trooper, we drove on more slowly, and it felt different. Pine, cypress, magnolia, and oak trees became individuals instead of a blur. Sunlight and shadows patterned the shoulder of the road. Fields between the patches of forest were no longer merely green, but a landscape tinted yellow by blooming mustard plants and red where the tops of sour weed waved in the breeze. I asked Johnny to stop so we could pick some sour weed. He bit into the stalk and said, "Wow, this stuff really is sour." Small amounts are supposed to be good for you; too much is toxic. Sour weed is something of a metaphor for life: what is good for you in moderation can kill you if you don't know when to stop.

At a bridge we slowed to see a "black-water" river, which refers to water stained the color of tea by the tannins from decaying leaves. Cypress trees lined its banks, limbs draped with beards of wispy gray Spanish moss. From time to time a farmhouse appeared, trucks parked in the front yard and clothes hanging on the line behind the house, reminding me of the farms of relatives I visited as a child. Some had magnolias planted, with their showy green leaves and white blossoms lending elegance to a modest box home.

We passed through small towns with maybe one traffic light in a commercial core that seemed barely to be hanging on to life. Empty tobacco warehouses and abandoned gas stations with pumps waiting for customers who would never come were all that were left of another time. All the visual stimuli charged my batteries for my visit home.

Later, I thought back to this ride and that state trooper. We never saw him coming or where he went, and I hardly ever remember a state trooper passing up a chance to write a ticket. More than an admonition for safe driving, his advice to slow down reminded me of the need to allow time for events to be absorbed, so that the distant memories of a life gone by can be recalled. Our unusual encounter presaged other unusual moments that occurred during this trip.

## Growing Up in Douglas

Slowing down also returned me to the rhythms of life in Douglas. By the time of my birth in 1941, my parents had established a life in that small southern town, but not the one they grew up expecting to have. The Great Depression had knocked the underpinnings out from under most family farms in the rural South, theirs included, and they had been forced to look elsewhere for a livelihood. After working jobs in places like Jacksonville and New Orleans, they managed to save enough money to buy the ice and coal plant in Douglas and return home. My sister, Phyllis, was born in New Orleans, but my brother, Ronnie, and I were born in Douglas.

We lived in a modest, pleasant home at 204 Grady Avenue on the west side of town. On my first visit in 2014, I checked, and the house was still there, with the addition of a carport. Sadly, the four longleaf pines that used to grace the front yard were gone. The family of one of my childhood friends, Francis "Frankie" Snow, lived three doors down from our home. Frankie still lives in that house, and I am fortunate to still have him as a friend. Educated as a botanist, he taught for more than forty years at South Georgia State College, located just south of town. His natural curiosity led him to become an expert on local Native American cultures and a first-rate archaeologist; he won the 2014 Crabtree Award from the Society for American Archaeology, which honors outstanding avocational archaeologists. He also has been a leader in conservation efforts, working to preserve ecologically sensitive land and protect endangered species. His name will come up more than a few times in this book because of the depth he added to my visits to my childhood home.

My mother and father were raised on farms to the north of Douglas in a county renamed in 1904 for Jefferson Davis, the president of the Confederacy. Farm families tend to be large, and those of my parents were no exception. Mom had nine brothers and sisters; Dad had eight. This fact alone gave me more than a few cousins in South Georgia. In addition, Cloughs have lived in these parts for more than 150 years, so it is not surprising that I have plenty of relatives nearby. As I explain to my citified friends, we don't marry our cousins down there, but there have been some close calls from time to time.

The author at eight years old, with his father and their Douglas home in the background.

During my childhood, on the farms of my relatives water came from wells, toilets were in outhouses, and while there were trucks and cars on the roads, mules still pulled the plows and tobacco sleds. Paved roads had yet to arrive, and bridges were wooden trestles that creaked and groaned when crossed. The cash crop was tobacco; farm incomes were small; and food came from fields, gardens, and rivers, not from stores. Jobs were either in agriculture or in work related to it, and poverty was not far from the surface for many. This way of life would end soon, with only dusty vestiges remaining in photos of faces long gone. Yet for all that was lacking, much was offered in the warmth of families, small-town values, and, for inquisitive boys, adventures near at hand.

For me, a boy with a wilding instinct, Douglas was a nearly ideal place to grow up. Life revolved around home, school, and the outdoors, where I spent my spare time exploring nearby swamps and pine barrens, fishing and hunting, and visiting farms owned by my relatives. Going barefoot was a given in the summer, and as winter turned into spring, we could not wait to leave our shoes at home when we left for school. We envied the poorer kids whose parents seemed to have a better understanding than our cautious families that walking on frosted grass did not cause colds.

Douglas did not have Yankee Stadium with Mickey Mantle, but we had the Douglas Trojans of the Class D Georgia State League. At the games, I was close enough to see the players spit tobacco juice after a good play. There was no natural history museum or zoo, but I had all the outdoors within a bike ride of my home. No orchestra or opera was anywhere to be found, but South Georgia College brought concerts and performances to the community. In addition, my sister was passionate about the piano (she went on to earn a master's degree from the Juilliard School). Classical music permeated our small house as she practiced and played her recordings of classical masterworks. The music of Chopin, Liszt, and Debussy left its mark on an impressionable boy.

Television had been invented, but we did not own one, so I listened to the radio, and my imagination was engrossed by tales of the Green Hornet and the Shadow. Douglas did have a movie house, the Martin Theater, where the Saturday double feature usually included at least one western, and weekly newsreels educated us about world issues.

Going to school in my day meant walking, heading east on Ward Street five or six blocks to the town center and then southeast a few blocks to the Douglas City Schools. The Clough Ice and Coal Company, owned by my father and mother, was located nearby, and I often stopped there on the way home to enjoy a shaved ice treat and say hello to my parents. Dad was the manager, and Mom was responsible for the accounting and human resource issues. Ice delivery to homes was one of the tricky parts of the business, since ice waits for no one.

The history I was taught at school was admittedly filtered through a southern perspective, and the menu was sparse. The Civil War was treated reverentially with little discussion of the complexity that lay behind it. In reality, the wiregrass country around Douglas with its sandy soils was no place for a *Gone with the Wind* Tara, and many of the hard-working people

The Clough Ice and Coal plant, owned by Dan and Bessie Clough, July 15, 1947.

who had lived there before the Civil War had harbored anti-big-planter sentiments. One legacy of the war was the hardship it brought both black and white citizens for decades to come.

To live in Douglas at the time of my boyhood was to live in the segregated South. Segregation was the rule in churches, schools, and hospitals, and politicians ran campaigns aimed at dividing potential voters along racial lines. At the Martin Theater, African American kids could only sit in the balcony while white kids sat below. We were separate but not equal.

I was fortunate that I never heard my parents utter a racial slur when I was growing up. My father, whose business relied on both black and white workers, always told me to respect people for what they did, not the color of their skin. As the years passed, I was impressed by the changes in Douglas and in Coffee County. Today the principal of Coffee County High School, Dr. Rowland Cummings, and the president of South Georgia State College, Dr. Ingrid Thompson Sellers, are African Americans, and their student bodies are reflective of the local population.

Douglas came to be where it is because of two roads: Peterson Avenue runs north and south; Ward Street runs east and west. Where they cross is the center of town. Peterson Avenue is U.S. 441, which during my childhood was the main road used by just about everybody on their way to or from Florida. Eventually an interstate replaced it as the major route to Florida, and I-75 bypassed Douglas.

When I was growing up, the principal institutions of the town were located around the intersection of Peterson and Ward, including the two funeral homes, schools, churches, law offices, the bank, the movie theater, and city and county government offices. A monument to the Confederate soldier, erected in 1911 by the United Daughters of the Confederacy, also stands next to the intersection.

Along Peterson and Ward toward the edges of town were the engines of commerce. To the north on Peterson lay the cattle auction house, which had somehow been allowed to locate adjacent to the town cemetery. The cemetery was large, having served as the final resting place for people from Douglas for more than 150 years. For most of this time, whites were buried in one part of the cemetery and black people in another. Separate in life, separate in burial. Many of my relatives are interred there, including my parents.

Also on the north side of town were the shops and roundhouse of the Georgia and Florida (G&F) Railroad, a short track that delivered the timber, cotton, and tobacco grown in the area to plants elsewhere to be converted into finished products for consumers. The decision by the executives

Downtown Douglas, Coffee County centennial celebration in 1954.

of the railroad in the early 1900s to locate in Douglas provided well-paying jobs to many people, some of whom were my relatives. The water needed for the G&F shops and steam engines came from a pond created by damming a creek and flooding a nearby swamp. The aptly named Shop Pond played an outsized role in my life as a child. Closer to town, the cotton gin was located on West Ward Street along the G&F tracks, so the cotton bales could be loaded easily on the freight cars.

South of town along Peterson were large tobacco warehouses, where farmers brought their crop in late August and September to be auctioned to the big tobacco companies. My mother would make boiled peanuts, which I sold for a nickel a bag to the farmers and their kids. When my peanuts were sold out, I ran water for the auctioneers and picked up tips from the men thirsty from talking at a speed beyond my comprehension.

A few blocks farther south was the campus of South Georgia College, now South Georgia State College, which had its own airport, built to teach students to be pilots and airport workers. During World War II the airport was chosen by the U.S. Army Air Force as a training base, and it was considerably upgraded to accommodate hundreds of pilots.

Douglas being a small town, all of the places of commerce were reachable by a kid on a bike such as me. With few distractions otherwise, warehouses, machine shops, and cotton gins offered opportunities for mischief, especially on the weekends when nobody was around. Because my mother and father both worked in their business, I was an original latchkey kid with plenty of time on my hands and an overactive imagination that often got me in trouble.

## Living Large in the Great Outdoors

Many of my outdoor pursuits as a boy involved water. Shop Pond was a favorite destination, and there were always boats around that we "borrowed" and poled or paddled into the deep recesses of the pond. The still surface of the tea-colored water would be broken by water snakes or the occasional water moccasin with its head high above its uniquely buoyant body, achieving a good speed by undulating to create forward motion. From time to time my friends and I accepted a dare to swim in the pond, but we were always on the lookout for snakes and ready to light out for the boat. We also had to keep an eye above, because moccasins would sun themselves along the tree limbs. Once, a few seconds after we had bumped our boat into a cypress tree, we heard leaves rustling and twigs breaking and realized we had dislodged a snake from up in the tree. Rather than wait for it to join us, we instinctively jumped into the water only to find that the falling snake had missed the boat and was now a fellow swimmer. The rush to get back into the boat was one of my fastest swims of all time.

The rich habitat of a swamp hosts many species besides snakes, providing a home for great blue herons, songbirds, skinks, frogs, and seldom-seen bobcats. This small alternative universe allowed boys with time to roam to imagine themselves as Native Americans or Tom Sawyer running from the law with Jim. We built huts and tree platforms, which brought out the future civil engineer in me as I organized teams to collect scrap lumber (filching it if necessary), along with nails and roof tiles. At night we lit corncob pipes filled with rabbit tobacco collected from the roadside on the way to the pond. Little did we know that the Native Americans who lived on the land before Europeans arrived called rabbit tobacco "life everlasting" because it was a medicinal herb. We enjoyed biting into the stalks of sour weed, which lit up the taste buds with a jolt of tart and tangy juice. Its shock was like the old Mennen Skin Bracer commercial in which a splash after a shave was followed by a slap to the face and a grateful "Thanks, I needed that."

Fishing trips with my father to the great Okefenokee Swamp were a treat, since he was usually busy with the demands of his work. The Okefenokee lies southeast of Douglas and just south of Waycross, Georgia—the town where my forebears first settled in the early 1800s and where my grandfather Forrest True Clough was born. My great-great-grandfather Cuyler Hilliard had a large timber operation near Waycross that reached into the Okefenokee. He was joined in his business by Jonathan Gilman Clough, my great-grandfather, who married Captain Hilliard's daughter, Isabelle.

On Okefenokee fishing trips we used a flat-bottom boat powered by a small outboard, navigating along narrow watery paths littered with water lilies until we arrived at open reaches where we caught sunfish, perch, bass, and catfish. Alligators trolled the black water of the swamp with only their eyes and the very tip of their snout above water to give away their presence. In the dark, their eyes blazed yellow-orange when we shined a flashlight across the water. At that time alligators were on their way to extinction, although I did not know it then. Fortunately, before these ancient creatures were lost, they were put on the endangered species list and have been brought back to a sustainable population.

Rivers played an important part in my South Georgia childhood, and their courses did not always take them where you might expect. Most people have trouble imagining a continental divide in flat South Georgia, but there it is. The Eastern Continental Divide stretches south from the North Georgia mountains and passes through Atlanta along Peachtree Street. As it continues through South Georgia, it passes to the west of Douglas and Coffee County, then swings rather sharply eastward for a stretch before rounding the northern end of the Okefenokee Swamp and turning south again along the eastern side of the swamp. As a result of this eastward curve, the Satilla River and Seventeen Mile Creek, both classic black waterways, flow east to the Atlantic Ocean. But another black waterway, the Suwannee River, made famous by Stephen Foster, rises farther east in the Okefenokee and flows to the Gulf of Mexico.

Although Douglas had a public swimming pool, we preferred the swimming holes found along the Satilla River. There was nothing like riding a rope swing off the top of a sand terrace above the river—arcing high out over the water, letting loose, and wrapping your arms tightly around your legs to form a cannonball. The goal was maximum impact and a rising plume of water that would have merited a ten had Olympic scorers been present. The more practical effect was to ensure that any nearby water snakes evacuated the vicinity.

Because Coffee County lies on the boundary between the upper coastal plain and the lower coastal plain, the region is the beneficiary of vibrant biodiversity. Swamps are only one player in an ecosystem that also includes sand terraces, bottomland forests, pine barrens, and open meadows, all serving as sites for my boyhood adventures. Songbirds, raptors, egrets, kingfishers, blue jays, crows, owls, hummingbirds, and red-headed, downy, and pileated woodpeckers filled the woods. In the more open pine barrens, where the longleaf pines and their symbiotic friend, wiregrass, thrived, gopher tortoises dug large, deep burrows, which they shared with dozens of other species, including nonpoisonous eastern indigo snakes. Both gopher tortoises and indigos have become endangered as this habitat has dwindled.

For me, the four longleaf pines that once graced our yard at 204 Grady Avenue were a connection to something that felt ancient and profound. Practically speaking, they provided shade for our family glider and its companion set of lawn chairs; during the evenings my family sat beneath our pines enjoying their whispering. True to their name, the needles, which grow in groups of three, could be up to eighteen inches long. When a hurricane veered inland, the high winds created a symphony swirling through the trees, whipping the tops back and forth. Running out into the warm sheeting rain of the hurricane beneath the longleaf pines left me feeling alive. The last night before we moved away, I stayed out as late as I could beneath the trees, hoping to take their music with me.

In addition to the longleaf pines and bald cypresses, South Georgia features southern live oaks, the state tree of Georgia. Their multiple trunks spread out in a splendid domed profile, and whiskers of Spanish moss hang in the low limbs. Live oaks earned their name because their leaves remain green year 'round. The sandy soils of the coastal plains also are home to the southern magnolia, which grows tall and wide with showy, forest-green leaves and great white lemony-smelling blossoms in the spring and summer. Where the southern magnolia grows wild in the bottomland forests, its companions are sweetgum, black gum, and water oak.

Pecan trees, common in the area, were also planted by farmers in orchards. These are not just any old pecan trees, but paper-shell pecans, producers of large nuts with shells that can be cracked by hand. My uncle Owen Walker lived farther out on the west side of town from us, but his home was reachable by bike. He had a large grove of paper-shell pecan trees that I loved to climb. The trick was getting high enough to reach the slender limbs, which swayed back and forth in the wind—a great ride for a kid otherwise earthbound. Finding a few pecans ready for the picking was an added treat.

It was not uncommon to find an arrowhead as I roamed the land as a boy. This did not make sense to me at the time, because there were no indigenous people in that part of Georgia. What I did not realize was that Native Americans had lived there for thirteen thousand or more years before Europeans arrived, and they had a rich, complex culture that white Americans are only now beginning to fully appreciate.

Of course, memories of my childhood would not be complete without mentioning insects. With all of those swamps comes a price to pay. Not surprisingly there were mosquitoes in the late spring and summer. And they did not just fly by; they came in attack squadrons. Douglas and other towns fought back with DDT fogger trucks, which trundled through the streets in the evening, trailing plumes of a poison now banned in the United States. I am not sure how much of a dent they made in the mosquito population, but they offered a diversion for bored kids. We delighted in running after the truck for as long as possible, almost invisible in the fog. That I am still here at my age is a wonder, given the amount of DDT I must have ingested.

Not biting but really aggravating were the gnats. Scientifically, gnats, a form of black fly, are known as *Liohippelates bishoppi*. They swarm around the faces of humans and animals, hoping to drink from the fluids we emit from our orifices. Despite possessing only a microscopic brain, gnats have a knack for finding humans, using their highly developed ability to detect the carbon dioxide we exhale.

## Departing and a Change of Plans

During my trip to Douglas in the spring of 2014, I looked for landmarks remembered. More than a few were gone, including the Georgia and Florida Railroad roundhouse, the tobacco warehouses, and my school. Still there, although sometimes in altered form, were my home on Grady Avenue, the old Clough Ice and Coal plant, the Confederate soldier monument, the Martin Theater, the cotton gin, and Shop Pond. The pond has shrunk in size from when I was a child, since its waters are no longer needed for the railroad works and steam engines.

More than a few of my cousins and relatives still live in and around Douglas, and I was able to spend time with them on that visit. I also was able to reminisce with childhood friends like Frankie Snow and Francis Lott, who have continued to reside in Douglas. Francis, a Georgia Tech alumnus like me, has been successful in life and given back to the town and community, helping it survive and thrive when other towns like it have dwindled and even disappeared.

My visit allowed memories to flood back in, but soon the time arrived to return to Savannah and fly back to Washington. Our plan was to drive by way of Augusta, where we were to visit the Hollywood Mound site, a Native American site where the Smithsonian had done archaeological work in the late 1800s.

That plan changed when a driving rainstorm hit Augusta and flooded the site. Instead, we drove with Frankie to the Douglas City Cemetery and found the gravesites of my parents and of the Snow family. My parents had been forced by financial circumstances to leave South Georgia two times in their lives, but this was home, and they made it their purpose to spend the ends of their lives and be buried there.

When we left Douglas, I suggested to Johnny Gibbons that we drive back through Waycross, which was only slightly out of our way to Savannah. I wanted to see the Kettle Creek Cemetery because I had been told that some of my relatives were buried there. With the help of Google Maps we located the Kettle Creek Congregational Methodist Church, founded in 1828 just outside of today's Waycross. The church got its name from a nearby waterway where a group of militia soldiers once camped. They were looking for a raiding party of Creeks who had attacked a settlement and wiped out a family. The militia left a kettle behind, hence the name of the creek.

Arriving at the church, we saw a surprisingly large cemetery. I realized that finding my relatives might be a challenge and suggested to Johnny that we begin by looking for the oldest headstones we could find. Spotting one darkened by age and lichens, I made out the inscription: "In the memory of / Col. John T. Clough / who departed this life / January 14, 1859 / aged 41 years / eleven months and nine days." Just like that, I had found my great-great-grandfather, who had moved from Northfield, New Hampshire, to Georgia in 1837. It was said he had come to find a warmer clime that would be kinder to his rheumatism, but cheap land, available because of the forced removal of the native Creek people, also had been at play in the decision. He earned the rank of colonel in the Georgia militia during the Second Seminole War.

Nearby I found the graves of other relatives, including Jonathan Gilman Clough, my great-grandfather, who had fought as a teenager in the Civil War and survived the conflict. His headstone reads, "In loving memory of our father / Jonathan Gilman Clough / December 8, 1845–November 26, 1907." Lying next to him is his wife, Isabelle Hilliard Clough, with a gravestone inscribed, "I have fought the good fight / I have finished my course / I have kept the faith / July 23, 1849–Jan. 21, 1937." Isabelle's father—another of my great-great-grandfathers—lies next to her in a grave marked by an elegant monument with the inscription "CAPT. CUYLER HILLIARD / BORN / Nov. 29, 1825 / DIED / Apr. 12, 1903 / REST IN PEACE." He earned his rank in the Twenty-Sixth Georgia Infantry, a unit assigned to the Army of Northern Virginia during the Civil War.

Around them were more Cloughs and Hilliards as well as other names I recognized from perusing my family tree—Millers, Sweats, and Hargreaveses. William Miller, who lies under the oldest gravestone in the cemetery, fought in the Revolutionary War and was one of the first settlers in the area.

As I pondered what I was seeing, my cell phone rang. It was my cousin John Clough, who said he was just checking up to see that I was okay on my travels home. The funny thing was, I didn't remember him ever calling me before to ask how I was doing. I told him I was fine and where I was. He said, "Oh, if you are there, you should go see where Captain Hilliard and Jonathan Gilman Clough once operated a store. The site is nearby on the grounds of a Baptist retirement home and is marked by a large live oak tree."

This was a coincidence similar to that state trooper a few days before. After all, I was supposed to be in Augusta looking at the gravesites of ancient Native Americans, not in Waycross looking at the gravesites of my relatives. There was no question we had to find the oak tree. We drove about a mile down the road to the Baptist retirement home, where a cautious guard at the entrance listened to my story. He said he didn't personally know anything about the tree, but it was right over there, and he pointed to a spot between two buildings. Sure enough, there it was, a force of nature with multiple trunks and a leafy canopy spreading out over an amount of space no ordinary tree could cover. I had never seen a live oak tree that big, and I could imagine it shading the patrons of a store and providing a respite from the heat of a South Georgia summer. I could envision the men whose graves I had just seen standing on the porch, going about the business of living.

After I returned home I learned that the big oak tree is the Georgia state champion, the largest of its species in the state, and is known as the "Village Sentinel." The Georgia Forestry Commission says "it stands 75 feet high with a limb spread of 155 feet and is said to be over 250 years old." In other words, this tree was alive when the Declaration of Independence was signed by three men from South Georgia: Lyman Hall, Button Gwinnett, and George Walton.

I stood there thinking of all that I would have missed if it had not rained in Augusta, if we had not stopped at the Kettle Creek Cemetery and looked for old headstones, and if John Clough had not called to ask how my trip was going. And then there was that state trooper telling us to slow down. As a scientist who believes everything has an explanation in facts, I was having second thoughts.

In fifth-century China, a philosopher called Lieh-tzu gathered ancient stories and musings into a collection. They are accessible to us in *Lieh-tzu: A Taoist Guide to Practical Living*, which says, "How we explain coincidences depends on how we see the world. Is everything connected, so that events create resonance like ripples across a net? Or do things merely co-occur and we give meaning to these co-occurrences based on our belief system? It is all in how you think."

As we headed to Savannah, all these coincidences swirled in my head. I saw a sign for a store selling boiled peanuts and asked Johnny to stop. Johnny had never eaten boiled peanuts, which are usually an acquired taste,

The "Village Sentinel" live oak at the Baptist Retirement Home, Waycross, Georgia, with the author in front of it. This was the location of the store owned by Cuyler Hilliard and Jonathan Gilman Clough.

but he politely said they tasted pretty good. I thought they were okay, but far from the fresh ones my mother had made for me.

In the course of writing this book, I made many other trips to my childhood home. As I discovered objects in the Smithsonian collections that related to South Georgia, the need grew to see things firsthand. The more I saw and learned, the more I needed to see. Each visit reinforced the last. The observations I made on my trips are now part of the narrative that unfolds in the chapters that follow.

CHAPTER TWO

# The Process of Discovery

Finding artifacts in the Smithsonian collections for my purpose—exploring South Georgia—required a framework for the discovery process. I had access to Smithsonian curators who know their subjects beyond my imagining, but they needed guidelines to know where to start.

My quest began with the notion that the most likely places to find artifacts would be in the natural history collections because birds, insects, reptiles, amphibians, and mammals are abundant in South Georgia. However, these creatures do not care a whit about human boundaries, which are artificial concepts. Animals and birds do not know when they have crossed the state line from Georgia to Alabama. The Suwannee River, which originates in the Okefenokee Swamp in Georgia and flows with ease down into Florida, has no knowledge of passing through two states.

So, rather than boundaries, I decided to use rules. Rules are guidelines, but they have exceptions and allow for judgment. In basketball, for example, the ball must go through the hoop to score. But how it gets there can vary. Some referees, for example, are known for "letting them play," which provides more flexibility before the ball goes up. That was the style of play needed for this project.

## Geographic Rules for the Search

State lines may not have been a perfect boundary for my search, but they are all over the maps we use, so the northern state line of Florida and the eastern state line of Alabama served as the southern and western limits for the search. A limited set of exceptions were allowed; they are noted below. As for critters, a particular species might live in Florida and Alabama as well as in Georgia or might migrate across state lines with the seasons. I included those that make their home at least part of the year in Georgia—especially those that came into the collections from the state.

Looking east for a control point, I came to the obvious boundary, the Atlantic Ocean. Abundant marine life, including mammals, such as dolphins and migratory whales, and many species of fish, reptiles, and crustaceans, is found just off the coast of Georgia. That marine life could have been the subject of a chapter of this book, because the Smithsonian has a collection of marine fossils from this area. But in the interest of avoiding too much mission creep, I stuck with the Atlantic coast as the eastern limit of South Georgia. Fortunately, this designation included all of Georgia's coastal ecosystem with its marshes and wetlands.

Looking north, I needed a rule that would divide South Georgia from North Georgia. One possibility was the gnat line, which nominally defines the area where these pesky insects live. Although there is some debate about where the gnat line starts, most agree that it cuts east-west across the state roughly where the city of Macon is located. Below it, the land tends to be sandy and dotted with swamps—the environment called home by gnats. They love standing water, and standing water is to South Georgia as glaciers are to Alaska.

Using a flying creature to determine a dividing line is not a very precise approach, but the real reason for the gnat's habitat comes down to geology. Georgia is formally divided into four geologic provinces beginning in the

The South Georgia gnat, *Liohippelates bishoppi*, magnified fifty times. National Museum of Natural History entomology collections.

The fall line in Georgia.

north with the Valley and Ridge and Blue Ridge, followed by the piedmont, and in the south the coastal plain. The coastal plain was formed over millions of years by repeated periods of ocean intrusion and retreat caused by climate change. As glaciers melted, sea levels rose, and the ocean moved inland into what is now Georgia. When the climate cooled and glacier formation resumed, sea levels dropped, and the ocean waters withdrew from the land. The departing waters deposited sandy sediments in their wake, which over time formed the coastal plain.

In Georgia, the northernmost boundary of the coastal plain is defined by the fall line, which is actually a zone about twenty miles wide that lies along a northeast-trending curve connecting Columbus on the western boundary of the state with Macon in the center and Augusta on the eastern boundary. When rivers and streams running south toward the Gulf of Mexico or the Atlantic Ocean leave the piedmont formation and encounter the coastal plain, they find softer soils that can be eroded more easily, creating the waterfalls that give this boundary its name. The waterfalls were used to good effect in the late 1800s and early 1900s as a source of cheap power generation, which helped entice textile mills from the north to move south. The so-called fall line cities—Columbus, Macon, Milledgeville, and Augusta—all were beneficiaries.

Below the fall line, the land becomes flatter and sandier, and the rivers begin to meander, their slow graceful movement only visible on the surface in swirling eddies around the roots and knees of cypress trees. Farther south there are places where the water runs out of energy and stops a while to await a surge from above before moving on. Such places are called swamps, and they are vibrant ecosystems—incubators of life that are at their best when humans are present the least. The habitats of the coastal plain lend themselves to flora and fauna not found north of the fall line. Alligators, for example, are found below the fall line but rarely above it. Snakes and insects, such as gnats, also are more abundant below the fall line than above it.

For my purposes, using the fall line as the northern boundary for the search was useful because it defined a special place in terms of habitat and geology. This was my last geographic rule for South Georgia.

## People Who Qualified for the Search

Beyond the geographic parameters was the question of what qualified an individual to be included in a book about South Georgia. It was obvious that anyone who had lived there all of his or her life would be in, but few people fit this criterion: it's a test even I do not pass. Instead, my rules on who could be included were pretty broad in order to open a wider and more interesting window on history. My first rule was that if you were born or died there, you qualified. Next, if you lived there for at least a few years, you were in. In this category, for example, was Eli Whitney, who lived in Savannah for two years, working as a tutor and being an inventor on the side. Finally, people who made history while in South Georgia qualified. In this category, Union general William Tecumseh Sherman did more than just squeak in. Also included under this rule was the president of the Confederate States of America, Jefferson Davis, who qualified despite only spending a few days in the area because his capture in South Georgia represented the end of the Confederacy.

With the rules in place, it was time to start searching the collections for artifacts that could be said to have a connection to South Georgia.

## The Smithsonian Collections

The Smithsonian Institution has twenty units that hold and manage collections, including the museums and galleries, the Smithsonian Archives, and the Smithsonian Institution Libraries. The holdings are among the largest in the world.

The collections are so extensive that they require large facilities outside Washington, D.C., to house them, with specialized sites for sculptures, paintings, photographs, automobiles, trains, insects, fishes, worms, fossils, Native American artifacts, and space suits. There are also large greenhouses containing the many plants and flowers used by the Smithsonian Gardens to beautify the institution's outdoor spaces and to provide flowers for museum visitors to enjoy. Among these are more than eight thousand varieties of orchids.

The numbers within the numbers are equally staggering. For example, there are 145 million scientific specimens, 10 million cultural objects, and 450,000 works of art. There are 350,000 bank notes, 2,200 live animals at the National Zoo, and two 1933 Saint-Gaudens double eagle gold coins. There is only one Hope Diamond.

Beyond the numbers, each item in each collection has a story. Every individual insect, of which there are tens of thousands, has been identified with a tag noting where it was collected, what species it is, who collected it, and when. I consider these labels a kind of ticket to immortality for the tiny creatures.

## The Early Days

The Smithsonian collections began before the institution was founded in 1846. Having seen what a national expedition could do after Lewis and Clark's voyage of discovery, President John Quincy Adams authorized an ambitious naval expedition in 1828 to circumnavigate the globe. The idea was to demonstrate to the world that the emerging democracy that was the United States could undertake a great initiative. The United States Exploring Expedition was launched in 1838 under the command of a young navy officer named Charles Wilkes. It consisted of a small armada of U.S. Navy ships whose primary purpose was to establish sailing routes for the

Bird collections, National Museum of Natural History.
Ornithologist Roxie Laybourne is in the center.

navy and for merchant ships. Wilkes also included eight scientists among his crew to collect cultural and scientific specimens. And collect they did. The expedition did not return until 1842, and it gathered more specimens and objects than had been collected in all three of the famous voyages by British sea captain James Cook combined.

On his return Wilkes personally ensured that the collected items were cataloged appropriately, but eventually the navy realized it had no use for most of the objects and no place to keep them. After a lot of give and take, the collections from the Wilkes expedition were given into the care of the newly minted Smithsonian in the 1850s.

Spencer Baird, the first assistant secretary of the Smithsonian and, later, its second secretary, was a zoologist, an ornithologist, and a driving force behind the early expansion of the collections of the Smithsonian. Baird deputized hundreds of people around the country to collect for him. In the 1850s he found more than forty willing volunteers in Georgia who had science or archaeology backgrounds or were willing to learn, and they sent specimens to build the growing Smithsonian science collections. With a diverse animal, fish, insect, and bird population, Georgia yielded a treasure trove for Baird.

President Theodore Roosevelt, an avid supporter of the Smithsonian, oversaw the gift that led to the creation of the first Smithsonian art museum. In 1906 his friend Charles Freer, an art collector, asked Roosevelt if he could contact the Smithsonian about a possible interest in his collection. It took more than a decade to complete the process, but the final result was the beautiful Freer Gallery of Art, which stands next to the Smithsonian Castle.

## Growing the Collections

Over time the collections expanded through multiple means. Individuals donated materials; curators collected to support their work; collections were gathered for new museums; and in some cases collections came with the acquisition of existing museums.

Since the 1950s, the Smithsonian has built the National Museum of American History, the National Air and Space Museum, the Udvar-Hazy Center of the National Air and Space Museum at Dulles Airport, the Hirshhorn Museum and Sculpture Garden (for contemporary art), the Anacostia Community Museum, the National Museum of the American

Indian, and, most recently, the National Museum of African American History and Culture. All of these museums have required collections.

In the 1960s the Smithsonian took over the old Patent Office Building, which was renovated to house the National Portrait Gallery and the Smithsonian American Art Museum. Similarly, the Renwick Gallery was acquired and renovated to house the growing collections of crafts and decorative arts of the Smithsonian American Art Museum. Not to be overlooked is the National Postal Museum, which opened in 1993 based on a joint agreement between the U.S. Postal Service and the Smithsonian. The Smithsonian was already the custodian for the National Philatelic Collection, but the postal service brought its own collections and continues to help build them.

In addition, over a number of years the Smithsonian accepted responsibility for the Cooper Hewitt Design Museum (New York City), the Archives of American Art, and the National Museum of African Art. Each of these entities came with collections.

On my watch as secretary, I was privileged to participate in the creation of the National Museum of African American History and Culture. Starting with no collections, this museum now has more than forty thousand objects and works of art. The collections tell the story of African Americans' contributions to our nation and also how our country has changed for the better.

Today, the Smithsonian only grows its collections when the need is clear. But history never stops being made: elections continue to be held, wars are fought, discoveries are made, new stamps are issued, advances are made in space exploration, and nature continues to surprise us. So the collections do expand, even as many artifacts are deaccessioned when they are no longer vital.

## Using the Collections

Exhibitions at the Smithsonian museums only display a tiny number of the institution's artifacts. Larger numbers of objects are loaned to other museums around the country for exhibition. Many of the artifacts not exhibited in museums can be viewed online as the result of a massive initiative by the Smithsonian to provide digital images of them. The collections are a resource for scholars, who come by the thousands to study them. Recently, the objects have found even greater relevance as researchers apply new techniques, such as DNA analysis, to see them in a different light and enrich our understanding.

During my tenure as secretary, the Smithsonian collections sometimes kept me up at night. I felt a great obligation to see to the care of these treasures on behalf of all Americans and all the people of the world. It has been hard to ensure that adequate funding is available year after year to conserve and preserve them. This need is ongoing, but because it is not glamorous or new, it seldom attracts the attention of donors.

I would like to think that my search for South Georgia artifacts illustrates a new way of using the collections. My project demonstrates how an individual can weave a set of time-based personal memories into a larger fabric that stretches over a broader space and a longer time frame. Of course, I was in a unique position that enabled me to call on the wide array of Smithsonian resources I needed for my quest. However, the digitization of the collections will allow anyone to explore them to their fullest. In the near future, everyone will be able to explore the Smithsonian's digital resources to make personal connections to their family, their history, our nation, or even the universe.

CHAPTER THREE

# Beginning with Fossils

## *How a Fossil Mystery Leads to a Story of Two Lost Eras*

The first stop in my search to find objects in the Smithsonian collections from my South Georgia boyhood home was an unlikely place: the Dinosaur Hall in the National Museum of Natural History. Since no fossil evidence from the Jurassic period has yet been unearthed from Georgia's geological strata, there are no dinosaur bones from South Georgia to be found at the Smithsonian. However, the creature I had come to see in the Dinosaur Hall was not a dinosaur.

Once I had set the parameters for my search, the biggest problem I faced was where to start. As the secretary of the Smithsonian Institution, I could have simply issued an SI-wide email asking for help, but that did not seem sporting. Instead, I turned to the sage advice of bank robber Willie Sutton who, when asked why he robbed banks, said, "Because that is where the money is." Applying Sutton's logic to the Smithsonian led me straight to the natural history collections because 94 percent of the objects and specimens in the collections are related to natural history.

I began with a call to Scott Wing, a paleobotanist in the Museum of Natural History, whose collections were the first I saw when I came to the Smithsonian. He and I had become friends, and he had invited me to Wyoming to participate in one of his summer digs, where he taught me to excavate fossils. Wing recommended that we talk to Brian Huber, who at the time headed the paleontology collections.

Huber was encouraging. He said he thought he could come up with a connection to South Georgia and suggested it would be useful to begin with a visit to the Dinosaur Hall. When we arrived, the hall was full of kids and parents awed by the huge fossils of creatures long extinct, which loomed over them. One girl shouted, "Mom, that's a big dinosaur!" Our destination, however, was a giant ground sloth.

When I came up with the notion of connecting my South Georgia home to the Smithsonian collections, I had no idea I would need to learn about giant ground sloths. But this venture proved to be perverse in that way. Connections, no matter how arcane, demanded to be followed, and the learning that resulted, even if it were not sought, was part of the process. I did not know it at the time, but I was on a journey during which a fossilized giant ground sloth and many other artifacts would lead me to a new understanding of myself and our world.

The first tip that an unexpected story was about to unfold was that despite being exhibited in the Dinosaur Hall, the giant ground sloth was a very different thing. Dinosaurs were reptiles and became extinct sixty-six million years ago. The giant ground sloth was a mammal, and while it also is extinct, it disappeared only about ten thousand years ago. It was clearly time for me to do some research.

It turns out that nobody even knew giant ground sloths existed until a fellow named Manuel Torres found one in 1788 in Argentina. The fossilized bones were sent to the Natural History Museum of Madrid where they were assembled to show what the creature might have looked like. It was big, as big as a grown elephant, and no one, including scientists, had ever seen anything like it before. In fact, there had been nothing like it in Europe or Asia because these unusual animals were native to the Americas.

It would take a contrarian to sort out the new creature, and he was a French scientist named Georges Cuvier. Cuvier had developed a theory that was controversial at the time: species could go extinct, some suddenly, and their existence could later be proven using fossils. Even though Cuvier had earned his spurs by careful work, most of his fellow scientists did not support his idea. When he saw drawings of the newly discovered Argentinean fossil, Cuvier concluded it fit his theory and was a species of giant ground sloth that had gone extinct. One of the keys to Cuvier's identification was the animal's big claws, which resembled those of the smaller sloths that still could be found in the tropical forests of South America. Many people disagreed with him, but he was eventually proved right. Cuvier named the fossil *Megatherium* ("great beast") *americanum* (for its home).

Soon after the discovery of the Argentinean *Megatherium*, giant ground sloths were given an unexpected boost in popularity when a large claw and some limb bones of another one were found in the United States. According to historian Silvio Bedini, they were given to an avid amateur paleontologist, who presented a paper in 1797 to the American Philosophical Society.

*Eremotherium* fossil in the Dinosaur Hall. National Museum of Natural History vertebrate paleontology collections. Kirk Johnson, the museum director, is with the author.

It was titled "A Memoir of the Discovery of Certain Bones of an Unknown Quadruped, of the Clawed Kind, in the Western Part of Virginia." The amateur paleontologist happened to be Thomas Jefferson, who was vice president of the United States at the time. Jefferson's specimen was smaller than the Argentinean *Megatherium* and from a different subspecies, but it was still a large sloth. It was given the formal name of *Megalonyx* ("large claw"), and later it was granted the type species name of *jeffersonii* in honor of the man who introduced it to the world. Jefferson must have been proud of this distinction, but his political opponents taunted him as "Mr. Mammoth" for spending time on his beloved fossil collection when they felt he should have been working on matters of state. Although Jefferson was ahead of his time in many ways, he was among those who did not believe in Cuvier's extinction theory. He even went so far as to advise Lewis and Clark to be on the lookout for live *Megalonyx* on their historic voyage of discovery.

## The Mammal Fossil in Dinosaur Hall and What We Did Not Know

Huber wanted my visit to the paleo collections in the Museum of Natural History to begin in the Dinosaur Hall so that I could first see the giant sloth as a completed whole. Then he took me into the paleo collections that are housed in the museum to see some of the "spare parts" that were used to determine what the sloth had looked like when it was alive. The giant sloth skeleton on display is actually only partly a real sloth, since it was constructed using skeletal remains that were incomplete. Plaster parts made to look like the real thing finished it off, and it is here that South Georgia enters the equation.

Leaving the Dinosaur Hall, we made our way through several floors containing hundreds of large collection cabinets; the dim halls are only fully lit when in use. We walked into a central area where paleontologist David Bohaska had arrayed a selection of bones on a metal table. Among the odd joints and leg bones was the lower jawbone of a large creature with molars about twenty times bigger than those of a human. The collection tags were yellowed with age and indicated the remains had been classified as those of a *Megatherium* and had been obtained from "Skiddaway" Island by Dr. J. P. Seriven. This fossil find was important to science not just because it was an intact jawbone of the creature, but also because it was the first to show that the *Megatherium* had existed in North America. (Hang on to this thought, because it turns out there is more to this story.)

While we were viewing the fossil bones, I noted that "Skiddaway" was probably Skidaway, an island that is almost a suburb of Savannah, Georgia. I had visited it several times as president of Georgia Tech, because it has a marine station that university scientists use as a base for research. I felt sure of my conclusion because the state park on the island has a small museum that features an exact copy of the giant ground sloth in the National Museum of Natural History.

I was enthused. On my very first visit to the museum collections, I had seen a fossil from South Georgia. And it was an important one, since this fossil was used to establish the presence of a species previously unknown in the United States. However, as would turn out to be the case more than a few times in my search, what seemed to be a done deal was not done at all. First, there was the word "Skiddaway" on the collection tag. Could it be more than a simple case of misspelling? Then, Huber told me that what was written on the collection tag as the species of the specimen reflected

*Eremotherium* fossil bones (originally identified as *Megatherium*) from Skidaway Island. National Museum of Natural History vertebrate paleontology collections. Paleontologists Dave Bohaska and Brian Huber are with the author.

*Eremotherium* fossil jawbone (originally identified as *Megatherium*) from Skidaway Island. The collector was incorrectly identified as Dr. J. P. Seriven. National Museum of Natural History vertebrate paleontology collections.

the state of the art at the time. More recently, changes had been made in the classification of giant ground sloths. I must not have been paying attention, because somehow I missed any mention of this news in the media. As a result, Huber said, the Georgia fossil was most likely an *Eremotherium*, not a *Megatherium* as the collector had thought.

Most people who wander into a museum to look at fossils for fun would have a hard time noticing any difference between *Eremo* and *Mega* sloths, but to experts there is a significant difference. It seems the two were pretty similar in size (i.e., big), but according to Darren Naish, "the former species

[*Eremo*] is characterized by a shallower maxilla with reduced hypsodonty of the upper teeth compared to the latter species [*Mega*]." I had no idea what "reduced hypsodonty" is, but I learned that the *Eremotherium* was the North American descendant of the South American *Megatherium*. The two began to separate into different species some three million years ago when a group of adventurous *Megas* moved north across the newly formed land bridge between North and South America, which later became known as the Isthmus of Panama.

This movement of species from South America to North America and vice versa is formally known among paleontologists as the Great American Biotic Interchange, but it was an unequal exchange. The creatures that headed south from North America were typically more successful than those going north, so what would become our giant sloth, the *Eremotherium*, was an exception. As to the South Georgia collector who misidentified the fossil remains in the 1800s, we can absolve him, because the difference between the two species was not understood until 1948.

When I reviewed the documentation in the fossil records with the help of the Smithsonian Archives, I found that the jawbone originally had been donated in 1842 to an organization called the National Institute for the Promotion of Science in Washington, D.C. The Smithsonian did not open its doors until 1846, but it soon eclipsed the National Institute, which folded in the 1850s and gave its collections, including the fossil from "Skiddaway" Island, to the Smithsonian.

To find out about the collector, I did an online search for J. P. Seriven and found a number of people with that name, but none seemed to fit the bill. Related names did keep popping up, however, namely Dr. J. P. Screven or Scriven. Regardless of the spelling, these references seemed to be about a man who had lived in Savannah about the same time as the fossil discovery.

With a little work I located a source in the Chatham County Archives that had been written in 1913 by William Harden about Dr. J. P. Screven. Harden wrote that Dr. James Proctor Screven, who was born in 1799 in Bluffton, South Carolina, came from a family with deep roots in the area. He had relatives who fought in the Revolutionary War, the War of 1812, and the Indian Wars of Andrew Jackson. Family members operated rice plantations in the area, but J. P. Screven was cut from a different cloth than most of his contemporaries, and he chose to attend medical school at the University of Pennsylvania. After receiving his degree in 1820, Screven was supported by his father for two years while he lived first in England and then in France to observe the medical practices in different countries.

While in Europe, he spent time studying geology and natural science as a matter of personal interest. It was an enlightened era when scientists were in high pursuit of discoveries. New developments were frequently announced, leading to improved understanding of mountain building, the effects of glaciation, and the evolution of species. After he returned to the United States, Screven set up a medical practice in 1822 in Savannah, but he kept up his interest in science and history.

A memoir written by William Hodgson in 1846 provided the details of Screven's involvement with fossils. Hodgson reported that Screven was a friend of another medical doctor in Savannah, John C. Habersham, who was an avid fan of fossils and antiquities. According to Hodgson, in 1823 Screven and Habersham were invited by a plantation owner named Stark to examine fossil bones that were exposed at low tide in a soil bank adjacent to a tidal pond on his property. Hodgson stated that the plantation was on "Skiddaway" Island, confirming my hypothesis that the spelling on the collection tag at the Smithsonian was an older form of today's Skidaway.

Screven and Habersham acquired a set of fossil bones from the plantation, and after Screven had studied them, he identified them as a species of *Megatherium*. He moved quickly, reporting his findings to the Georgia Medical Society in 1823. Poor Habersham may have gotten the short end of the stick in this business, since it would turn out that he was by far the more committed of the two to paleontology. Regardless, the paper by Screven provided the documentation that he was the donor of the fossils to the National Institute for the Promotion of Science; "Seriven" was a misspelling on the specimen tag.

Screven's interests soon moved away from fossils and toward his medical practice and in 1835 to full-time work on his inherited South Carolina and Georgia landholdings and rice plantations. His plantations produced a large fortune for him, but rather than living a life of leisure, he moved to downtown Savannah and set about doing all he could to improve the city. Serving as an alderman and eventually mayor, he is credited with developing a clean water system, a gas supply system, and the public schools of Savannah. He died in 1859.

We don't know much about what Screven did with the fossil bones after he identified them as *Megatherium* in 1823, but in 1842 he presented drawings of them to a meeting of the National Institute for the Promotion of Science in Washington, D.C. Shortly thereafter he donated the fossils to the National Institute, a gift that I confirmed through the Smithsonian Archives with the help of Smithsonian historian Pam Henson. She also

## Lessons of the Megafauna

My first foray into the paleontological collections seeking connections to South Georgia ended successfully. I not only learned about fossils but also uncovered stories, long buried, about the people who discovered them and the times in which they lived.

Although the Smithsonian collections contain only one South Georgia fossil of note, it is an important one that established the presence in North America of a new species of giant ground sloth, once thought to exist only in South America. That fossil also laid the groundwork for fossil discoveries made later in South Georgia, which identified other new species and documented the existence of the remarkable animals that populated South Georgia and other parts of North America before going extinct more than ten thousand years ago.

I was pleased that in a small way my research had provided the Smithsonian with information to sort out the mystery surrounding the fossils I saw and give credit to individuals where credit was due. As to my own edification, I had known relatively little about the era of megafauna in South Georgia, and I came away with an appreciation for a time of change in which hundreds of species were lost while another species, humans, adapted and thrived.

That animals like giant sloths, bisons, giant beavers (*Casteroides ohioensis*), mammoths, mastodons, and horses once roamed my home of South Georgia twelve thousand years ago is amazing. As one author said, in the long view of geologic time, we just barely missed them.

While working on this book, I gave a keynote address at the annual meeting of the American Society of Mammalogists in Jacksonville, Florida. The agenda included a field trip to White Oak, a conservation center located just south of the Georgia-Florida line that features large African animals. While we could see large African animals now living in a place close to my home, the large native animals that had once lived there would have been even more fascinating.

CHAPTER FOUR

# Book Learning

Not long after my encounter with the giant ground sloth in the Dinosaur Hall, I made my way back to the National Museum of Natural History for a search that I expected to be a bit less dramatic. Deep in the museum, I walked into an elegant wood-paneled room that is part of the Smithsonian Libraries. In this room is a large refrigerated container. Its low temperatures help preserve rare books, many of which predate the founding of the Smithsonian in 1846. Some contain intricate colored drawings of what America's earliest non-indigenous naturalists saw during their journeys.

My interest was in three of these books, each written in the 1700s about the natural history of the southeastern colonies of Great Britain and the Spanish colony of Florida. They describe the place that existed before European immigrants changed it forever.

In the 1700s Native Americans still held most of the land in the southern colonies. Although English settlers were arriving in waves, they were largely confined to the coastal regions. The tinder for the fire of conflict was building, however, and war would come in time. In the interim the constant jockeying for position often led to open hostilities. Despite the hazards posed by the colonial conditions and the conflicts between peoples for the land, a few non-indigenous naturalists were interested enough in the subtropical environment of the Southeast to take on the challenges. These men became legends in the fields of anthropology, ornithology, botany, and entomology. Among them were Mark Catesby, John Abbot, and William Bartram. All three spent years exploring the Southeast and recording what they saw. Today collectors pay $100,000 or more for first editions of their books. Since I don't have that kind of money, it was fortunate that the earliest editions of their books are housed in the rare book collections of the Smithsonian Libraries.

Curator Leslie Overstreet had arrayed the treasured documents on tables for me to review. Unlike today's books, most of these are large folios that require two hands to open; they are filled with detailed illustrations. The pages are tinged yellow by age and lined with words that speak of another time.

Catesby, who was senior to Bartram and Abbot, was the first to undertake an expedition, covering the years 1722 to 1726. He published his work in two volumes under the title *The Natural History of Carolina, Florida, and the Bahama Islands* (1731, 1743). A supplement was published in 1747, just two years before he died. The second volume opened with this excerpt from a dedication written to Her Royal Highness the Princess of Wales:

> Some of the most wonderful productions of America have now the honour to lay themselves at the feet of your royal highness; Things new and strange to the inhabitants of this side of the globe, and which must necessarily excite in the minds of infinitely less discernment and piety than that of your royal highness exalted and religious sentiment of the power, wisdom, and goodness of that Almighty Being who has not only formed the universe but adorned every part of it with an inconceivable variety of beauty and magnificence.

Beauty and magnificence indeed. His book and those of Abbot and Bartram contain illustrations that almost leap off the page. Bright in their detail and nuanced by the rendering of light and shadow, the sketches bring birds, butterflies, flowers, and animals to life in a way no photograph can. These books tell a story of the natural history of the southern colonies more than two hundred years before my childhood memories were made. Although theirs was a different era, I was surprised that many of the species were familiar to me. Nevertheless, they also described vast forests and creatures that disappeared before my time. The naturalists' books and the historical narrative that was the context for them provided a perspective for my journey through the Smithsonian collections looking for South Georgia.

## Mark Catesby

Catesby journeyed through the wilds of what is now the Southeast United States at a time when Georgia did not yet exist. The colony was not established until 1733 by James Edward Oglethorpe, seven years after Catesby had returned to England. Its southernmost boundary with Florida would not be fixed until Andrew Jackson's Indian Wars were concluded in the 1800s. So, when we read Catesby's *Natural History of Carolina, Florida, and*

The "parrot of Carolina" and the "cypress of America." From Mark Catesby, *The Natural History of Carolina, Florida, and the Bahama Islands*, vol. 1 (1731). Smithsonian Libraries rare book collections.

*the Bahama Islands*, Georgia as we know it today is included in the southern lands of Carolina and the northern regions of Florida.

Catesby particularly focused on birds, many of which I knew as a child, such as the eastern bluebird (*Sialia sialis*), the bald eagle (*Haliaeetus leucocephalus*), the red-headed woodpecker (*Melanerpes erythrocephalus*), and the flamboyant pileated woodpecker (*Hylatomus pileatus*). In other cases, his paintings were of birds I knew, but their names were of his time. The "mock bird" and the "fox colored thrush" would eventually become known as the mockingbird (*Mimus polyglottos*) and the brown thrasher (*Toxostoma rufum*), which was named Georgia's state bird in 1935.

His book is also a reminder that in the more than two hundred years between his visit to South Georgia and my childhood, species were lost. His illustrations of the "parrot of Carolina" and the "pigeon of passage" show the Carolina parakeet (*Conuropsis carolinensis*) and the passenger pigeon (*Ectopistes migratorius*), beautiful species once so numerous they flew in great flocks. The last known specimens of each species died in captivity in 1914 and 1918, respectively. The fate of the great "Lord God bird,"

The ivory-billed woodpecker and the willow oak. From Mark Catesby, *The Natural History of Carolina, Florida, and the Bahama Islands*, vol. 1 (1731). Smithsonian Libraries rare book collections.

the ivory-billed woodpecker (*Campephilus principalis*) shown so lovingly in Catesby's book, presently is unknown since rare claims of sightings have yet to be confirmed.

After completing his travels in the New World in 1726, Catesby returned to England, never to revisit the land that had so captured his imagination. It would take him twenty-one years to finish publishing his work, which formed the foundation for the naturalists who followed him.

## John Abbot

Abbot, the least known of the three naturalists, was born in England and was expected by his family to become a lawyer. However, his love of the outdoors and his skill as a painter and illustrator led him to a career as a naturalist. In 1773 he moved to America, settling first in Virginia. During the Revolutionary War he moved to Georgia and lived the rest of his life in the countryside near Savannah. A first-rate artist, he focused on butterflies, moths, birds, plants, and small mammals. Although his illustrations were widely used because of their high quality, his name is attached only to one publication, *The Natural History of the Rarer Lepidopterous Insects of Georgia*, whose author was James Edward Smith. The book was published in 1797 and contains 104 images by Abbot.

Abbot's illustrations of butterflies include many I saw as a child, among them the painted lady (*Vanessa virginiensis*) and the Georgia state butterfly, the eastern tiger swallowtail (*Papilio glaucus*). Abbot passed away in 1840, having garnered little of the acclaim of Catesby or Bartram, because he had no publications in his own name.

## William Bartram

Bartram's ancestors migrated to the colonies in the 1600s, and more than a century later his father, John Bartram, became known as America's foremost biologist and horticulturalist. John Bartram traveled from his home in Philadelphia throughout the colonies to make collections and identify species. In 1765 he was named the king's botanist for North America, which provided an annual stipend to support his work. John Bartram was close to Benjamin Franklin, and the two participated in the establishment of the American Philosophical Society. Its name reflected the fact that at the time scientists who studied the physical world were referred to as natural philosophers.

Although William Bartram evinced early talent as an artist and botanist, it was not clear that he would become a chip off the old block. He struggled with motivation for years, and by 1764 had become a trading store operator in Cape Fear, North Carolina, but with little success. A turning point in his life came in 1765, when his father was charged with undertaking a botanical expedition to Georgia and Florida and asked young William to join him. The two made important discoveries of new species and set the stage for other naturalists, who would build on their findings. But even after the expedition, William Bartram refused to return to Philadelphia to follow in his father's footsteps. He attempted instead to become a planter in Georgia, only to fail yet again. He was rescued from his business misadventures by a benefactor, Dr. John Fothergill, who offered him funding to spend two years in Georgia and Florida making drawings of the wildlife and plants. In 1773 William set off on this journey, which lasted not two years but four. It changed his life forever and at the same time provided us with a lasting record of the natural history of Georgia and surrounding states.

Perhaps because of his eclectic life to that point, Bartram brought a sensibility to his work that others did not have. His interests were wide-ranging and included a fascination not only with the natural world, but also with the lives of the European migrants and Native Americans he met along the way. His work was complicated by the start of the Revolutionary War, which overlapped with the last two years of his expedition. During this time he dodged British troops and their Native allies and managed to avoid capture.

Bartram arrived in Charleston, South Carolina, in early 1773 and set about preparing for his first year of work. Most of it would be in and near coastal Georgia, all of which lies within what I have defined as South Georgia. With limited European settlements in the interior of Georgia at the time, the only trip on which Bartram penetrated into the interior during his first year was when he traveled from the coast up the Altamaha River. On this journey, he came close to the Forks, or the confluence of the Oconee and Ocmulgee rivers, which form the Altamaha. The Forks lies in the northeastern corner of Jeff Davis County, where my mother and father spent their childhoods and where many of my relatives have lived.

Most of Bartram's second year was devoted to Florida, but in the third year he traveled from Charleston up the Savannah River and then eventually southwest through Georgia on his way to the Gulf of Mexico. This

"Mico Chlucco the Long Warrior, or King of the Siminoles." From William Bartram, *Travels* (1791). Smithsonian Libraries rare book collections.

leg of his journey led him across the Oconee and Ocmulgee rivers well above their junction at the Forks. Crossing the Ocmulgee, he saw what he called the "Oakmulge Fields," where he noted the great mounds and terraces that had been built by Native Americans. These were the works of the Mississippian culture—people who had come to the area from the west in 800 CE. Bartram knew the people he met as Creeks (also called Muscogees). They were the survivors of the diseases that had ravaged their ancestors as a result of contact with the Spanish explorer Hernando de Soto and his expedition, which made its way across South Georgia in 1540. Today we know the "Oakmulge Fields" as the Ocmulgee National Monument, located just outside of Macon, Georgia.

Bartram completed his journey through the southeastern region in 1777, and then he returned to Philadelphia to write his opus. He was in no hurry to publish, and his book did not appear until 1791. Titled *Travels through North and South Carolina, Georgia, East and West Florida, the Cherokee Country, the Extensive Territories of the Muscogulges or Creek Confederacy, and the Country of the Choctaws*, the book surprised Bartram by becoming what we would call today a best seller. Europeans were particularly enamored of it, and Bartram was lauded for both his scholarship and his writing style. Because of the demand, his book was reprinted in 1792, just one year after it was first published.

*Travels* was said to inspire the work of poets like William Wordsworth and Samuel Coleridge, and later Bartram was referred to as America's first Henry David Thoreau. While Bartram's writing may seem in want of editing, his observations were keen, and he succeeded in telling of the beauty found in pristine streams and rivers, forests with abundant old-growth timber, and wildlife not yet diminished by overhunting.

Bartram found a way to combine scientific descriptions of plants, trees, insects, birds, and animals with his delight in the lushness of the semitropical world he had discovered and his wonder at how different it was from the manicured world of his childhood. A reader senses that he knew his principal job was to document the species he saw, but he seemed impatient to get this task out of the way so he could express his enthusiasm about seeing so many new things.

### *The Coastal Plain of South Georgia*

Bartram's first encounter in 1773 with the sandy terraces of the coastal plain and its great longleaf pine forests caused him to understand that this was a place of wonder:

> We now rise a bank of considerable height, which runs nearly parallel to the coast... when we find ourselves on the entrance of a vast plain, generally level, which extends west sixty or seventy miles.... This plain is mostly a forest of the great long leaved pine (P. palustris Linn.), the earth covered with grass, interspersed with an infinite variety of herbaceous plants, and embellished with savannahs, always green, sparkling with ponds of water and ornamented with clumps of evergreen, and other trees and shrubs.

Thus Bartram depicted the forests of South Georgia of almost 250 years ago, which were densely populated by the longleaf pine, a tree that creates its own ecosystem of open savannas. In this ecosystem was a highly diverse network of birds, reptiles, mammals, and grasses that could be found nowhere else. Where ground conditions were more accommodating for other species, Bartram noted the spectacular magnolia trees and flowering shrubs like the *Gordonia*. The *Gordonia*, an unusual shrub with beautiful white gardenia-like blossoms, is a close relative to another rare species Bartram and his father saw on their expedition to Georgia. On his second trip Bartram collected seeds and specimens for later study. The plant turned out to be a new species, named *Franklinia alatamaha* for Bartram's friend Benjamin Franklin and the river on which it was discovered. Bartram expected to find the Franklin tree, as it is popularly known, in other places, but it turned out to have existed only in one small grove near Fort Barrington and nowhere else. Today it does not exist at all in the wild, believed to have fallen victim to cotton planters. However, because Bartram collected the seeds, which were provided to horticulturists, it still grows in many gardens in the United States.

Much of what Bartram saw changed after the land was opened for farming and after logging began in earnest in the 1800s and early 1900s. The great forests of longleaf pine described by Bartram disappeared as the trees were sought for heart-of-pine lumber, first for local consumption and later for shipment overseas. Along with the longleaf pine (*Pinus palustris*), the stately cypress tree (*Taxodium distichum*) was harvested for its wood, which is not only beautiful but also long lasting. As these trees were cut down, no attempt was made to replant them because faster-growing non-native species were preferred. When the trees were lost, the symbiotic ecosystems they supported also disappeared.

In 1867 there were still extensive longleaf pine forests, which the budding naturalist John Muir encountered on reaching South Georgia during his first journey into a natural wilderness. His journal, *A Thousand-Mile Walk to the Gulf*, records these observations:

> September 30. Between Thomson and Augusta I found many new and beautiful grasses, tall gerardias, liatris, club mosses, etc. Here, too, is the northern limit of the remarkable long-leafed pine, a tree from sixty to seventy feet in height, from twenty to thirty inches in diameter, with leaves ten to fifteen inches long, in dense radiant masses at the ends of the naked branches. The wood is strong, hard, and very resinous.
>
> October 5. In "pine barrens" most of the day. Low, level, sandy tracts; the pines wide apart; the sunny spaces between full of beautiful abounding grasses, liatris, long, wand-like solidago, saw palmetos, etc., covering the ground in garden style. Here I sauntered in delightful freedom, meeting none of the cat-clawed vines, or shrubs, of the alluvial bottoms.

Both Bartram and Muir recognized the beauty of the longleaf pine and the ecosystem that made it special, but it took later research to show that the tree is an ancient species with a unique defense mechanism developed over millions of years of evolution. It uses fire as its friend, possessing an ability to withstand flames and high temperatures while competitive species succumb and are removed. Understanding the role of fire is a key to the restoration and health of the longleaf pine, something Native Americans had learned and used for generations as they reinvigorated the forests through controlled burns.

After Muir passed through Georgia and moved on to California, where his fame grew as a naturalist, steam power was brought to South Georgia, increasing the ability of timber harvesters to cut and transport trees, such as the longleaf pine and the cypress. A number of my relatives from Waycross, including my great-great-grandfather Cuyler Hilliard and my great-grandfather Jonathan Gilman Clough, were in the lumber business in the 1800s. By the early 1900s, however, the days of logging longleaf and cypress were numbered, because few trees were left to cut down.

Today the longleaf pine is found in less than 2 percent of its previous range, and species that are dependent on the longleaf ecosystem, including the red-cockaded woodpecker (*Picoides borealis*), the eastern indigo snake (*Drymarchon couperi*), and the gopher tortoise (*Gopherus polyphemus*), are on the federal endangered species list. Fortunately, private citizens; nonprofits, such as the Nature Conservancy and the Longleaf Alliance; government agencies, such as the Georgia Department of Natural Resources; and large landholders, including the Ichauway Plantation and the Department of Defense's Fort Benning and Fort Stewart, are working together to restore these beautiful trees and the endangered species that thrive where the trees do well.

### *"Animal Productions"*

Bartram also was charged with identifying the animal and insect species he encountered, and he did, but never with quite the gusto his botanical training instilled in him for his favorite plant species. After allotting many pages to flora on the coastal plain, he commented:

> As for the animal productions, they are the same which originally inhabited this part of North America, except such as have been affrighted away since the invasion of the Europeans. The buffalo (urus) once so very numerous, is not at this day to be seen in this part of the country; there are but few elks, and those only in the Appalachian Mountains. The dreaded and formidable rattle-snake is yet too common, and a variety of other serpents abound, particularly that admirable creature the glass-snake: I saw a very large and beautiful one, a little distance from our camp. The alligator, a species of crocodile, abounds in the rivers and swamps near the sea coast, but is not seen above Augusta. Bears, tygers, wolves, and wild cats (felis cauda truncata) are numerous enough; and there is a very great variety of papilio (butterflies) and phalena (moths), many of which are admirably beautiful, as well as other insects of infinite variety.

That Bartram saw wolves and "tygers" (cougars and panthers) is a statement of the times, since all large predators were eventually eliminated from Georgia by European settlers. It is said that panthers or cougars are seen today in South Georgia from time to time. If so, they have come north from Florida, where a hundred or so have managed to survive into our age.

Later in his book Bartram mentioned other animal species he saw in South Georgia, including one of my favorites, the fox squirrel (*Sciurus niger*):

> Here are several species of squirrels (sciurus) peculiar to the lower countries . . . and some of them are very beautiful creatures.
>
> The great black fox squirrel is above two feet in length from nose to the end of the tail, which for about two inches is milk white, as are the ears and nose. The red fox squirrel is of the same size and form, of a light reddish brown upper side, and white underside, the ears and tip end of the tail white.
>
> The grey fox squirrel is rather larger than either of the foregoing; their belly white, as are the ears, nose and tip of the tail. These three seem to be varieties of the same species.

### *The Rivers*

Out of necessity Bartram traveled mainly by rivers. He understood their importance to the coastal plain and explained how the courses of the Altamaha River and its tributaries were intertwined:

> It has its source in the Cherokee mountains [and from] . . . there it descends through the hilly country, with all of its collateral branches for two hundred and fifty miles and then enters the flat plain country, by the name of Oakmulge [Ocmulgee]; thence meandering on one hundred and fifty miles, it is joined by the Ocone [Oconee], which likewise heads in the lower ridges of the mountains. After this confluence, having gained a vast acquisition of waters, it assumes the name of Alatamaha [Altamaha], when it becomes a large majestic river.

On making camp, Bartram often felt a keen sense of being part of nature. An experience with a storm on the lower Altamaha in late 1773 inspired these words:

> Nature now weary, I resigned myself to rest; the night passed over; the cool dews of the morning awoke me; my fire burnt low; the blue smoke scarce rose above the moistened embers; all was gloomy; the late starry skies, now overcast by thick clouds, warned me to rise and be going. The livid purple clouds thickened on the frowning brows of the morning; the tumultuous winds from the east now exerted their power. O peaceful Alatamaha! Gentle by nature! How thou wert ruffled! Thy wavy surface disfigured every object, presenting them obscurely to the light, and they at length totally disappeared, whilst the furious winds and sweeping rains bent the lofty groves, and prostrated the quaking grass, driving the affrighted creatures to their dens and caves.
>
> The tempest now relaxed, its impetus being spent, and a calm serenity gradually took place. By noon the clouds broke away, the blue sky appeared, the fulgid sun-beams spread abroad their animating light and the steady western wind resumed his peaceful reign.

Passages like these set Bartram apart from other naturalists and inspired the romantic in readers. In the novel *Cold Mountain* by Charles Frazier and the movie of the same name, the protagonist, W. P. Inman, sustains himself through the brutal battles of the Civil War by clinging to a picture of his beloved Ada and a worn copy of Bartram's *Travels*. After he is grievously wounded and decides to desert the army, an old pistol, the picture of Ada, and Bartram's book are the treasured items Inman carries with him on the long, tortuous journey home. This speaks to the reverence held for Bartram's book even years after it was published. Since the late twentieth century, hikers have been able to walk the Bartram Trail, which follows the route Bartram used in his journeys, allowing us to see the natural heritage he so lovingly described.

## Natural Beauty Lost to Time

I did not go into the Smithsonian Libraries with a feeling of excitement about the rare book collections. I knew there were likely useful things to learn, but there were no specimens to see or fossils to touch. However, when I opened the oversized folios of the rare books, I found those remarkable illustrations of birds, trees, reptiles, and butterflies—made more than two hundred years ago—to be evocative. Like so many others, I was enchanted by these images.

In a practical sense my visit to the Smithsonian Libraries provided connections between the Smithsonian collections and South Georgia, as I had hoped, but there was more. I came to understand Bartram's lyrical way with words and realized that if you suspended your critical self, you could get swept away by them. His vivid descriptions of the great expanses of longleaf pines now gone, of the beauty of birds such as the extinct Carolina parakeet, and of animals such as wolves, which we will never see again in South Georgia, left me with a sense of loss.

In reading Bartram's *Travels* and the books by Catesby and Abbot, I gained an understanding of what South Georgia was like before my family arrived in the 1800s and how much of the wondrous ecosystem those authors saw has been lost. Now I could compare their descriptions to my childhood memories and be better prepared to appreciate what had changed and what remained, which we should seek to save for posterity.

and then a reddish one farther down the road. They forage on the ground and hop as much as run. Their antics and quick glances make it seem as if they know you are watching, and they are providing entertainment for you.

Helgen accommodated my request and invited me to a large collection room on the floor above his office. He spread his arms to encompass dozens of cabinets and said, "Here are the fox squirrels." Helgen explained that there were so many because fox squirrels are found in one form or another up and down the East Coast and into Canada. Their range reaches west to the foothills of the Rockies. As we looked at cabinet after cabinet of fox squirrels, Helgen described the variations of the species across different geographic regions. Finally, we walked slowly through cabinets labeled by state—Maryland, Virginia, North and South Carolina. In the Georgia section a tray had been pulled out with specimens collected in South Georgia, resplendent in coats of black, reddish brown, gray, and blond.

Helgen said that while the individual color variations were found elsewhere, the only place they were seen together was South Georgia. He noted that the fox squirrel was particularly well adapted for ecosystems such as the pine forests of South Georgia. Although hunted by some, their population is relatively stable and will continue to be if enough suitable habitat remains for them. I hope these whimsical creatures, in all their color variations, will continue to be found in South Georgia for generations to come.

Fox squirrels from South Georgia. National Museum of Natural History mammals collections.

On a later trip to one of the Smithsonian collection centers in Maryland, I extended my mammal count for South Georgia when I saw the skeletons of dozens of opossums, bobcats, and black bears (*Ursus americanus*), most of which had been collected in the Okefenokee Swamp. The swamp was on its way to being eliminated in the early 1900s by logging and the draining of water for agriculture. Fortunately, thanks to a few determined naturalists, it was protected as the Okefenokee National Wildlife Refuge in 1937. Today, it serves as home to a wide variety of species that otherwise would be endangered.

## Sustaining the Story

After my visits to the mammal collections, I reflected that on the surface the mammal populations of South Georgia seem to be in good shape. Regardless of the radical changes in the ecosystem over time due to cutting down trees and the introduction of large-scale agriculture, mammals, particularly small ones, have been able to adapt and move to safe havens. The Wildlife Division of the Georgia Department of Natural Resources, hunting associations, foundations, and a few committed individuals have helped save some species and stabilize others. We owe them a debt, but it is a never-ending process as the human population increases, habitat is threatened, and forces like climate change generate challenges.

The story for large mammals is quite different. Predators like the cougar and wolf, common in Bartram's time 250 years ago, are gone. Even Georgia's deer, which are plentiful today, were almost wiped out in the 1950s by overhunting and loss of habitat. In my childhood, deer were rare. To the credit of the Georgia Wildlife Division, hunters, landowners, and environmentalists, the deer population and that of the black bear, which was almost wiped out as well, have made comebacks and now have reached stable levels. Both of these stories provide a foreshadowing of the future, when active intervention will be needed if we are to keep the beauty and diversity of what is left of our natural environment.

CHAPTER SIX

# Birds and the Stories They Tell of the People Who Collect Them

Birds have fascinated people, inspiring thoughts of higher beings and magic and stimulating poets to write. Henry Wadsworth Longfellow: "Do you ne'er think what wondrous beings these?" Walt Whitman: "Skirting the river road . . . / Skyward in air a sudden muffled sound, the dalliance of the eagles, / . . . Upward again on slow-firm pinions slanting, their separate diverse flight, / She hers, he his, pursuing." John Edward Smallshaw: "A curlew knew me too as I sat here, but / it flew away holding the secrets that it would not say."

For most of my adult life I have carried around a worn copy of "The Wild Swans at Coole" by William Butler Yeats in my briefcase. I am not sure why, but when I am on a trip far from home or frustrated by the events of a day, reading this poem sets my mind at ease. It reminds me that nature and one of its most beautiful creatures, the wild swan, transcend our lowly human travails.

> Under the October twilight the water
> Mirrors a still sky;
> Upon the brimming water among the stones
> Are nine and fifty swans.
>
> The nineteenth autumn has come upon me
> Since I first made my count;
> I saw, before I had well finished,
> All suddenly mount
> And scatter wheeling in great broken rings
> Upon their clamorous wings.

Birds and their flight seem to free us from the weight that ties us to the ground. We love them, paint them, study them, and collect them.

There are some ten thousand species of birds, so it takes a big collection to represent them. Welcome to the Smithsonian Institution, where more than six hundred thousand bird specimens are under the care of the aptly named chief curator Carla Dove. Fortunately for me, she is one of the top scientists of the Smithsonian and a friend who was willing to help in my quest.

The bird collections at the Smithsonian are used by scientists from around the world to better understand the characteristics of species and their evolution over time. Beyond assistance with issues of basic science, Dove and her colleagues also are called upon to help with practical problems. For example, each year there are about eight thousand bird strikes—instances in which a bird hits an aircraft, civilian or military, in flight and at times damages the plane. Dove and her staff help to identify the birds who do this and provide advice on avoiding similar impacts in the future, saving birds, planes, and people in the process.

## Recollections of South Georgia Birds

My search of the bird collections did not rise to a life-threatening level nor call for chemical or biological analyses. Instead, the bird collections opened windows into my past, evoking childhood memories and, more surprisingly, posing questions about the South and its complicated place in our nation's history.

Bird species are well represented in Georgia because the state provides diverse habitats and ample food sources. The northern part offers mountains and hardwood forests, while the southern part is home to mixed pine and hardwood forests, swamps, marshes, and coastal ecosystems. In South Georgia, the protected habitat of the great Okefenokee Swamp alone has 230 species of birds.

Perhaps because birds are beloved by so many, they are a well-studied lot, and I can do little justice to them in this brief chapter. The most comprehensive work on the birds of Georgia is a 746-page tome appropriately titled *Georgia Birds*, written by Thomas D. Burleigh, a naturalist who lived part of his life in the state and taught at the University of Georgia. Burleigh collected bird specimens for his own purposes and for institutions, including the Smithsonian.

Not being an expert who can cite examples of rare birds, my childhood memories are of the usual suspects: northern cardinals (*Cardinalis cardinalis*), blue jays (*Cyanocitta cristata*), eastern bluebirds (*Sialia sialis*), mockingbirds (*Mimus polyglottos*), a variety of woodpeckers, American robins (*Turdus migratorius)*, brown thrashers (*Toxostoma rufum*), great blue herons (*Ardea herodias*), and so on. Hunting was part of the passage from boyhood to manhood where I grew up, and our quarry back then included bobwhite quail (*Colinus virginianus*) and mourning dove (*Zenaida macroura*), both hunted in the late fall and early winter after the crops had been harvested.

Today, upscale magazines depict quail hunters in expensive togs riding horses across groomed plantations. My experiences as a boy were much less elegant but perhaps more of and about nature. Rising early, we dressed in worn jeans, old jackets, and boots warmed by the dying embers of last night's fire in the fireplace. Our breath fogged in the cold air as we walked along the brush line next to the summer's cornfields, watching the bird dog work: its nose near the ground, then in the air, searching for the scent of a covey. The quail held still on the ground, hiding, their colors blending with the fall grasses. When the dog went on point, we would move ahead in unison to flush the covey, ready to click off the safety on the gun when the birds took flight.

Quail often wait until the last minute to leave their cover, but when they decide to go the entire covey lifts off, wings whirring, with each bird heading off in an unpredictable pattern. Although I tried to stay calm, my pulse would pound, and only instinct would enable me to single out one quail for a shot. I was never the best at it, because others just had more ability, but a quail hunt was all a boy could desire on a cold day otherwise to be spent indoors. What remains of those days is the .410 side-by-side, double-barrel shotgun my father gave me, which I have kept for no other reason than to look at it from time to time and allow the memories to seep back into my consciousness.

The population of bobwhite quail is now on the decline because of loss of habitat and its vulnerability to a variety of predators, which find the ground-dwelling bird and its eggs easy prey. A coalition of hunters, landowners, conservationists, and state wildlife organizations is working to help maintain the viability of the quail population.

## The Bird Collections

As was the case with the mammals, the sheer scope and complexity of the Smithsonian collection of birds presented the challenge of paring down my search to a manageable size. I began by asking if specimens of birds

I remembered as a boy were among the collections and if any had been collected from South Georgia. Keeping it simple, I asked about the brown thrasher, which is the state bird; the robin, which alerted me to the coming of spring; two of my favorites among the woodpecker family, the downy and pileated woodpeckers (*Picoides pubescens* and *Hylatomus pileatus*); and my hunting quarry, the mourning dove and the bobwhite quail, which is the state game bird.

On my visits to the bird collections, Dove and her colleague Chris Milensky showed me specimens from South Georgia. One seemed never enough, and dozens of each species were arrayed in specimen cabinets—brown thrashers, robins, downy and pileated woodpeckers, and mourning doves. As we looked over the groups of birds, I was struck by how many different people had collected them over such a wide range of years. Some specimens dated back to the 1840s, sent to the Smithsonian by volunteers who had been recruited by Spencer Baird when he was the assistant secretary of the institution. Others were more recent, with the most prolific collector being the indefatigable Thomas Burleigh. I was informed that Burleigh had donated more than twenty thousand birds to the Smithsonian during his long career, with more than a few of them from Georgia.

Among the birds contributed by Burleigh from South Georgia were four brown thrashers collected over the period 1936–1946 from various localities. It came as a surprise to me to learn that the brown thrasher is a member of the family Mimidae, which is Latin for "mimic." The more widely recognized members of this family are mockingbirds and catbirds (*Dumetella carolinensis*), known for their ability to mimic the songs of almost any bird or other sounds they might hear. My friend the brown thrasher turns out to be the best of the lot, having a repertoire of more than a thousand different songs and a strong, clear voice that is hard to miss. The brown thrasher spends most of its time on the ground, using its long, curved beak to work through leaf litter and the upper surface of the soil to uncover insects, nuts, and worms. Its subtle brown coloring, as evidenced by the specimens from the collections, is an adaptation to blend into its habitat and make it hard to see.

My initial search was successful for all the birds I had inquired about with one notable exception, the bobwhite quail. Although the collections contain many specimens of this species, none came from South Georgia. I was surprised, since Georgia's quail are prized as a quarry for hunts, especially in southwestern Georgia around Albany and Thomasville.

Four brown thrashers collected from 1936 to 1946 in South Georgia by T. D. Burleigh. National Museum of Natural History bird collections.

From time to time, largely for sentimental reasons, I accept invitations from friends to participate in a quail hunt, and Dove made me promise that if I were to go again I would collect some of the birds for the Smithsonian. She made her request more pointed by noting that five of the former secretaries of the Smithsonian had provided bird specimens for the collections. Although I observed that the previous secretarial contributors to the bird collections were ornithologists by training, I know I will remain under a cloud with Dove until I make good on my promise.

## Anatomy of a Comeback

Despite all of the time I spent roaming the woods of South Georgia, one large bird I never saw was a wild turkey (*Meleagris gallopavo*). Indigenous to the Americas, this bird was once common and a source of food for the Native Americans and early European settlers of South Georgia. Its numbers steadily declined in the late 1800s and early 1900s due to overhunting and loss of habitat, and it was on its way to extinction. Fortunately there has been a decided turnaround, and the population is now estimated to exceed six million. Today when I visit South Georgia, these remarkable birds can be found on farmlands and in forests everywhere.

The story behind the rebound of the wild turkey population began with the passage of the Federal Aid in Wildlife Restoration Act (commonly known as the Pittman-Robertson Act) in 1937. It directed that the excise tax on the sale of guns and ammunition be allocated to the states to be used for wildlife restoration. This act provided the seed funds for scientists working for state wildlife agencies to join forces with committed hunters and conservationists and lay the groundwork that saved our native turkeys.

It was late in my research for this book when I asked Chris Milensky about turkeys in the Smithsonian bird collections. He quickly provided documentation showing seven turkeys in the collections from Georgia, six of them from South Georgia. That in itself was good news, but the bonus was that the collector of five of the specimens was one of the most renowned conservationists to emerge from the South, Herbert L. Stoddard. Stoddard established his reputation in 1931 when he wrote a landmark book about the bobwhite quail, and he went on to develop broad conservation techniques for all the species living in longleaf pine forests. Stoddard provided turkeys to the Smithsonian between 1938 and 1950; two of the specimens came from his own home, Sherwood Plantation.

## Lost to Time: The Less Fortunate Species

The survival story for some South Georgia bird species is not as positive as that of the wild turkey. Of course, extinction is part of nature. On average, species on Earth survive between five hundred thousand and a million years before going extinct. But when humans enter the equation, the odds of an early exit go up. Such was the case with South Georgia's megafauna, which went extinct over the course of several thousand years after the first arrival of humans in the New World.

Humans changed the extinction equation a second time when Europeans poured in and took over the land. In his descriptions of South Georgia wildlife in the early 1700s, naturalist Mark Catesby reported seeing flocks of Carolina parakeets and passenger pigeons, and he documented the existence of the magnificent ivory-billed woodpecker (*Campephilus principalis*), the so-called Lord God bird. Turns out I was born too late to see any of these birds. The last known passenger pigeon and the last Carolina parakeet died in 1914 and 1918, respectively. Ironically, Martha, the last passenger pigeon, and Uncas, the last Carolina parakeet, died in the same cage in the Cincinnati Zoo. They exist today as mounted specimens in the Smithsonian bird collections. Other passenger pigeons and Carolina parakeets are also

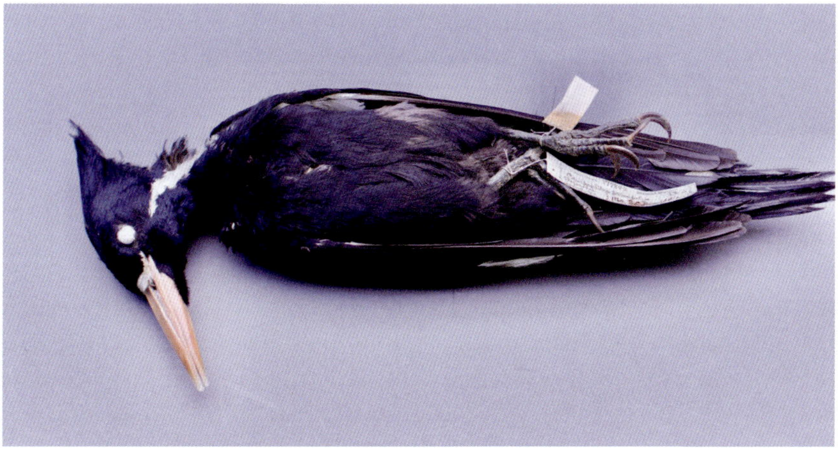

Ivory-billed woodpecker, female, collected by S. C. Dinwiddie in 1860, Okefenokee Swamp. National Museum of Natural History bird collections.

in the collections, and several of the passenger pigeons are from South Georgia.

The ivory-billed woodpecker is thought to be extinct. Although a sighting was reported in 2004, it has not been confirmed. I have seen the pileated woodpecker in South Georgia, and this smaller relative of the ivory-billed woodpecker is itself a remarkable sight. But the closest I suspect I will ever get to the ivory-billed woodpecker is seeing a number of specimens in the bird collections of the Smithsonian. Two of them came from South Georgia near or in the Okefenokee Swamp. The most intact specimen is a female collected in the Okefenokee in 1860.

Looking at the specimens of these birds in the collections, one can only think of how beautiful they were and how diminished our lives are because we will never see them fly or hear their songs. Beyond that sense of loss is the question, "how could we let this happen?"

## The LeConte Specimens

As I looked at one of the passenger pigeons collected in South Georgia, I discovered that the specimen tag listed the name of LeConte as the collector. This name had cropped up when I visited the mammal collections and saw a bat whose tag referred to the LeConte plantation in Liberty County, Georgia. Liberty County is located south of Savannah, roughly due east of Douglas. The tags on the bat and the passenger pigeon reminded me of a story I had heard that an entire bird collection had been donated to the Smithsonian by someone named LeConte. This story was intriguing because it offered the possibility of finding more specimens in the collections from South Georgia and perhaps learning about the collector as well.

I asked Milensky and Dove if they could find additional evidence of bird specimens linked to a LeConte. Sure enough, they were able to show me that close to a hundred specimens had been entered in the logbooks with the donor name of LeConte, and the location of most of the specimens was cited as either Liberty County or a town named Riceboro, which is in the county. The collection logs also showed a variety of appellations for donors named LeConte: J. LeConte, Jos. LeConte, Prof. Jos. LeConte, Prof. J. LeConte, and Prof. J. L. LeConte. Of these the most common was J. LeConte. The dates for the specimens ranged from the 1840s to the 1850s with two exceptions from Prof. Jos. LeConte: a donation in 1870 of the skeletal remains of a California condor (*Gymnogyps californianus*) and a donation from the Falkland Islands in 1872 of three penguins—two macaroni (*Eudyptes chrysolophus*) and one emperor (*Aptenodytes forsteri*)—plus one blue-eyed imperial shag (*Leucocarbo atriceps*), a cormorant.

As I considered the information at hand, there were loose ends. Why the variations in the collector names? Why did all of the birds before 1860 come from Georgia and then the few after 1870 from California and the far-away Falkland Islands? Also, among the specimens listed in the log were two bobwhite quail, but there were no bobwhite quail specimens from South Georgia to be found in the collections.

At that point, all I knew for sure was that one or more people with the name of LeConte had donated many birds to the Smithsonian and that most of these bird specimens, but not all, had come from South Georgia.

## A Family Passion for Science and Early Connections to the Smithsonian

I was fortunate at this point to have my friend Pam Henson mention the name of Lester Stephens. Dr. Stephens is a distinguished emeritus member of the history faculty at the University of Georgia and has written extensively about a group of scientists from the southern coastal states who became leaders in their fields before the Civil War. In most cases the Civil War cut short their careers, and the momentum they had created in the South for scientific discovery was lost. Among the people Stephens studied was Joseph LeConte, and he wrote the definitive biography of him, *Joseph*

*LeConte: Gentle Prophet of Evolution*, a copy of which is in the Smithsonian Libraries. In time I managed to visit Stephens at his home in Athens and interview him. With his insights and some additional information discovered in the Smithsonian Archives, the LeConte story and its connection to the Smithsonian bird collections began to emerge. Joseph LeConte in particular played a central role in the bird collections and in the explanation of how California came to be part of the story.

Joseph LeConte was the grandson of John Eatton LeConte, a man of scientific sensibilities who lived in New York but bought a rice plantation near the Georgia coast as an investment in the 1760s. Named Woodmanston Plantation, it was close to the town of Midway, south of Savannah. At the time, rice plantations in coastal Georgia were successful because of easy access to water, the presence of low-lying bottomland, and the unpaid labor of enslaved people.

When John Eatton LeConte died, his two sons, Louis and John Eatton Jr., split his property, with John Eatton Jr. choosing land in New York and Louis, who loved being surrounded by nature, acquiring Woodmanston Plantation. Louis not only sought to operate the plantation for a profit, but also took time to indulge his interests in science, including creating a bountiful botanical garden with flowers he had modified by crossbreeding. Louis and his wife, Anne Quarterman, had seven children. All were talented, but it was the middle two sons, John and his younger brother Joseph, who channeled their inherited love of science into careers.

Growing up on the plantation, John and Joseph developed a love of the natural world. Joseph was the more outgoing of the two, choosing to study geology and natural history, which called for fieldwork, while John chose mathematics and physics as his focus. Both earned undergraduate degrees from the University of Georgia, John in 1838 and Joseph in 1841, and then received medical degrees from a prestigious university in New York City. They returned to Georgia, set up medical practices in different cities, and started their families, but both were pulled toward science.

John was appointed professor of chemistry and natural philosophy at the University of Georgia in 1846. Joseph left his medical practice in 1850 to study geology at Harvard in the new Lawrence Scientific School with Professor Louis Agassiz, the most famous scientist in the United States at the time. Joseph then joined his brother on the faculty of the University of Georgia, but both soon moved to South Carolina College (which became the University of South Carolina) because of disagreements with the president of the University of Georgia. Back in New York, John Eatton LeConte

Portrait of Joseph LeConte by Louis O. Lussier, 1876. National Portrait Gallery collections.

Portrait of John LeConte by Louis O. Lussier, 1877. National Portrait Gallery collections.

Jr., the uncle of John and Joseph, established his name working in the U.S. Army Corps of Topographical Engineers. While he was recognized for his work as a topographical engineer, he also became a noted naturalist and is particularly recognized as an expert on turtles. His son, John Lawrence LeConte, became famous as America's most prominent entomologist, classifying more than half of the beetles known to exist in the United States at the time.

All four LeConte scientists—John, Joseph, John Eatton Jr., and John Lawrence—had at least a passing interest in birds throughout their lives. While John and Joseph grew up at Woodmanston, their uncle, John Eatton Jr., and their cousin John Lawrence spent time there as well. As young men, John Lawrence and Joseph took a long field trip together in 1844 to visit the Great Lakes region. On this trip they identified a species of sparrow that summers in the Great Lakes and winters in southeast Georgia. This bird was ultimately named LeConte's sparrow (*Ammodramus leconteii*).

Over time the brothers John and Joseph documented the birds of Georgia, and they collaborated on the first comprehensive listing of the state's birds, which appeared in George White's 1849 *Statistics of the State of Georgia*. The two brothers, particularly Joseph, also established a growing scientific collection of bird specimens.

All four LeConte scientists had linkages to the Smithsonian Institution, some of which had begun before the Smithsonian was created. Central to these connections was Spencer Baird, a young ornithologist who was related to the New York LeContes by marriage. Joseph had become friends with Baird while working on his medical degree in New York City in the early 1840s. By 1850, Joseph LeConte was beginning his studies at Harvard, and Spencer Baird was beginning his thirty-seven-year tenure at the Smithsonian, the last nine years as the second secretary of the institution. Baird and the LeContes remained friends, and Joseph sent bird specimens to Baird for the Smithsonian collections. In 1858 Joseph sent a letter to Baird confirming that he was turning over the remainder of his collection to the Smithsonian. His donations spanned a period of time during which he made career changes, explaining why the titles associated with "J. LeConte" in the collection logs changed.

We will probably never know the exact number of birds Joseph LeConte gave to the Smithsonian because fewer birds remain in the collections than are recorded in the collection logs. In its early days, the Smithsonian had a practice of loaning specimens to scientists on request, but the institution was not always able to ensure that the specimens came back. The two bobwhite quail listed among the donations from Joseph LeConte were probably lost in this way.

### *John and Joseph LeConte: Unexpected Connections*

The donation of Joseph's entire bird collection to the Smithsonian was not the end of the LeConte story. The next chapter actually strengthened their connection to the Smithsonian, to its collections, and to me. During the 1850s John and Joseph saw their reputations for their scientific work grow, and they came to the attention of other scientists. In 1856 Joseph was asked by Smithsonian secretary Joseph Henry to deliver a series of lectures in the auditorium of the Smithsonian Castle, which at that time was the only building on the National Mall. (As secretary, my office was in the castle.) As noted by Joseph LeConte in his autobiography, "In December of 1856, I therefore gave three lectures on coal and three on coral reefs. They were delivered unwritten, but those on coal were afterward written out and published in the Smithsonian Report for 1857." LeConte's lectures contributed to his reputation as a budding scientist of great potential.

Henry's appreciation for the scientific work of the LeConte brothers was high, and he asked them in 1858 to undertake work for the Smithsonian.

Joseph's work involved geologic mapping, and John's focused on the development of a table of constants for "nature and art." But this working relationship drew to a close with the onset of the Civil War. The LeConte brothers' experience in chemistry was sought for the development of munitions, and they were drawn into service for the Confederate cause. Joseph was even given the rank of major.

The war ended badly for the LeContes. William Tecumseh Sherman's army destroyed much of the Woodmanston Plantation, in which John and Joseph still had partial ownership. On the campus of the University of South Carolina, Union troops spared the buildings, but the scientific papers and records of both brothers were burned, and Joseph thought his career might be at an end. After the war, the university had limited resources and struggled to reach a basic operational level. The LeContes chafed under the working conditions of the Reconstruction era, which included serving under African Americans who were appointed to administrative positions by the Republican Party. While they complained that the black administrators lacked background and competence, without question the matter of race also played a role, since they now found themselves working for people they considered their inferiors.

They began to search for a place to begin again, and their Smithsonian friend, Joseph Henry, came to the rescue. Henry had many contacts and knew some of the members of a group working to create a new state university in California, which would become the University of California, Berkeley—the school at which I would earn my PhD in 1969.

In a letter dated August 13, 1866, to a Dr. Fitzgerald in San Francisco, Henry wrote:

> I have written to S. J. Butterworth, Esq, warmly recommending the appointment of Professors Joseph and John LeConte of the University of South Carolina as suitable candidates for the chairs of Natural History and Chemistry for the new University of California. I am intimately acquainted with these gentlemen and am free to say that I do not think it possible for you to obtain two more learned or accomplished teachers in the line of the branches I have mentioned than these two brothers. They are not only well acquainted with the latest discoveries in science but are capable of original investigations.

John LeConte became not only the first faculty member at Berkeley but also its acting president during its formation. He later served as its president and as a distinguished physics professor. Joseph LeConte followed his brother, moving to California with his family and bringing along their older sister Jane and her children. Both Joseph and John became widely known for their original work in science, and both were elected to the National Academy of Sciences.

Joseph made original contributions in geology and was a master teacher; in 1878 he published *Elements of Geology: A Textbook for Colleges and the General Public*, which was used in the United States and around the world. An original copy of this book is in the Smithsonian Libraries. He wrote four other science books, including *Religion and Science* (1873) and *Outlines of the Comparative Physiology and Morphology of Animals* (1900). Both are held in the Smithsonian Libraries. He was the first professional geologist to give his endorsement to John Muir's proposal that glaciers had formed the spectacular features in Yosemite Valley. His long friendship with Muir led to a commitment to conservation, and he joined with Muir to found the Sierra Club, which he served as a director. Joseph's love of Yosemite Park led him there on dozens of trips; he died at Yosemite in 1901.

Joseph and John LeConte's involvement with the Smithsonian did not end with their employment at the University of California. Both of them maintained a correspondence with Joseph Henry until his death in 1878, and they remained friends with Spencer Baird for the rest of their lives. Perhaps this ongoing friendship, combined with nostalgia for the old days, explains why Joseph LeConte continued to send bird specimens to the Smithsonian collections from California, including the California condor, the penguins, and the blue-eyed cormorant that I saw on my visits to the bird collections.

The many contributions of the LeContes were recognized by the University of California in the naming of a new physics building in their honor, LeConte Hall. A street near campus was named LeConte Avenue. In honor of Joseph's commitment to conservation and his contributions to geology, his name is found in the Sierra Nevada on a mountain, a canyon, a falls, a divide, a point, and a lodge. Also bearing the LeConte name are a glacier in Alaska, Mount LeConte in the Great Smoky Mountains National Park, LeConte Hall at the University of Georgia, and LeConte College, a science building at the University of South Carolina.

In today's world, place names are increasingly under scrutiny because of concern about linkages to slavery and racism. The LeConte name on facilities has been questioned by the Sierra Club and on the Berkeley campus. Without doubt, John and Joseph owned slaves and considered African Americans inferior to whites. Joseph expressed this position in a book, *The Race Problem in the South* (1892). He seems to have overcome biases against new scientific theories like evolution, and even moved to a more tenable position on women's rights over time, but he was unable to see past his racial prejudice. How history will ultimately judge John and Joseph LeConte is yet to be decided.

The two men can be seen in photographs housed in the Smithsonian collections. Photographs of each, taken in midlife, are in the National Portrait Gallery. A third photo of Joseph, taken in his office at the University of California in 1898 when he was seventy-five, captures him in a pensive mood—a man who had lived a full life.

Although I am a fellow South Georgian, my small orbit intersected that of John and Joseph LeConte in California. When I enrolled at the University of California, Berkeley, in 1966 to study for my PhD in civil engineering, I knew nothing of the LeConte brothers, even though I went to class in LeConte Hall and parked on LeConte Avenue. At that time I was hoping to put my past behind me because in California the principal impression of the South was of reactionary governors, resistance to civil rights, and whites who opposed progress. I, however, was optimistic about the future and set on learning all I could from a world-class faculty. Never was a better decision made by a young man who had strayed from his South Georgia roots. My experiences at Berkeley laid the groundwork for a career that I never thought possible.

Many years later, Berkeley was kind enough to honor me as a distinguished alumnus of the College of Engineering. Because of this connection and because one of my former PhD students is now a faculty member there, I return to campus from time to time. Whenever I visit, I enjoy seeing the places where I was a student, but now I also carry the knowledge of two brothers from South Georgia who moved there after the Civil War and rebuilt their lives. I think about their legacies, in which scientific accomplishments are mixed with a history tinged by racism.

## A Visit to the Woodmanston Plantation

I eventually realized I had to visit the site of the Woodmanston Plantation. It was no longer a place of myth, but a reality that had lodged itself in my consciousness. My friend Frankie Snow helped me connect with Terry and Caroline Harrington, who hosted my visit. For many years Terry had taught with Frankie at South Georgia College in Douglas. (We later figured out that Terry and I are related. I seem to be related to just about everybody in southeastern Georgia.) Caroline was a member of the trust for the LeConte-Woodmanston Plantation and Botanical Gardens, a group working to partially restore the site so visitors can understand its place in history.

On a sunny February day in 2014 I drove south from Savannah on I-95 with my Smithsonian colleague Johnny Gibbons and as instructed took exit 76 and headed east toward the town of Midway, a familiar name from my LeConte readings. From there the directions took us along back roads, paved at first but changing to sand and shells toward the end. All the while the cypress trees and saw palmettos crowded the roadway, and the black water of a swamp glistened through the overgrowth on either side of the car. Eventually we reached an opening in the forest, and we had arrived at the plantation. A sign explained that the threat from the mosquito population was "low," which was a good thing since the other possibilities listed were "high" and "dangerous."

Terry and Caroline Harrington were waiting for us and escorted us to grounds that had been cleared—grounds that once were home to the plantation as well as the famed botanical gardens of Louis LeConte. The site is on higher ground than the surrounding swamp and is graced by large oaks and cypress trees festooned with Spanish moss. Although no LeConte

Caroline Harrington and the author at LeConte's Woodmanston Plantation and Botanical Gardens near Midway, Georgia, March 2014.

A LeConte Norman red camellia at LeConte's Woodmanston Plantation and Botanical Gardens.

family buildings remain, a replica of a slave cabin has been built by descendants of the enslaved people who worked on the plantation, some of whom live nearby in Midway. Caroline told us that many of the descendants have the surname LeCount—a variation of the name of the family that once owned their ancestors.

To the side of the plantation, volunteers have developed the beginnings of a botanical garden with flowers and shrubs like those that were planted and hybridized by Louis LeConte, including his camellias: the charming tricolor, the graceful Ave Maria, and the blazing LeConte Norman red. A trail leads visitors through the surrounding wetlands, allowing them to see the works and canals that were once used for growing rice. Walking through the grounds and wetlands of the Woodmanston Plantation, one can easily imagine why a young Joseph LeConte was inspired to learn about the flora and fauna that provide its beauty.

Before we left, I walked a little by myself and closed my eyes, listening as the breezes whispered through the trees. It was almost as if I could hear voices from the past. I was glad I had come, and I felt a sense of closure to this chapter of my journey, which had first begun in the mammal collections of the Smithsonian, where I saw a bat collected in 1831 from a place known as the LeConte plantation.

CHAPTER SEVEN

# Amphibians, Reptiles, and a Detour into My Family History

For a young boy growing up in South Georgia, reptiles were a force to be reckoned with. Any mention of snakes in particular brings back vivid childhood memories, some of which spook me to this day.

On one occasion I was in the swamp behind Shop Pond just outside Douglas, looking for a rope to tie up a boat my friends and I had found. On a dry patch of land covered by leaf litter I saw what I took to be a coiled rope and reached for it—only to hear a distinctive rattle. In an instant I realized that my hand was no more than four inches from a big eastern diamondback rattlesnake (*Crotalus adamanteus*) looking straight at me with glittering black eyes and a tongue that darted back and forth. I froze as the snake and I eyeballed each other to see who would make the next move. Then slowly, ever so slowly, I began to pull back my exposed extremity.

For some reason the rattler held back, and when my hand was what I considered to be a safe distance away, I jumped backward. I felt grateful to the snake for not biting me when it could have and to God for letting me live. As I watched, the snake decided this little tête-à-tête with a human had gone on long enough and slithered away, its colors quickly making it all but invisible. I could not help but wonder how many other times I had been close to a rattlesnake and did not know it.

Over the course of my boyhood, I came into contact with all types of reptiles and amphibians and did my best to let them decide that I was not a threat or that I was too fast to be caught. That I am still here also means I had a generous helping of luck, particularly given that I went barefoot a lot and loved to swim. Any body of water would do, whether a river, lake, swamp, or cow pond. My swimming stroke was more breaststroke than crawl, because I had to be sure to get things that were in the water out of the way before I passed through.

## The Amphibian Collections

Given the reality that a good number of reptiles and amphibians join me in calling South Georgia home and that an entire section of the Smithsonian's natural history collections is devoted to amphibians and reptiles, I had high expectations about the possibility of making a connection. The prospects for my search were enhanced by my host, Jeremy Jacobs, a scientist in the National Museum of Natural History who has worked in one capacity or another at the Smithsonian for more than forty years and whose specialty is amphibians and reptiles. To add icing to the cake, when Jacobs was earning his PhD at the University of Maryland, he did some of his field research in South Georgia's Okefenokee Swamp, so he knew the territory.

There are two places to see amphibians and reptiles in the collections of the National Museum of Natural History: in the museum itself and in the Smithsonian's vast collection center in Suitland, Maryland. My visit was to the collection center, which was built to provide a space where many collections could be maintained in conditions allowing for improved access and environmental control.

After I checked in at the visitor's desk at Suitland, Jacobs took me past a series of large concourses and rooms with labels like "Fish" and "Mammals." In time we came to "Amphibians and Reptiles." Jacobs began by telling me that no one interested in the amphibians and reptiles of South Georgia should be without the bible on the subject, *Amphibians and Reptiles of Georgia* by Jensen, Camp, Gibbons, and Elliott. When Jacobs said he was giving me his personal copy, I knew I was now obligated to spend more than a few of my spare evenings reading. It was a good problem to have. I came away with a new appreciation about these remarkable animals, which are critical to our natural diversity.

Amphibians were abundant in the streams, ponds, and forests that were part of my wanderings and explorations as a boy. Salamanders could be found by turning over a rotting log, and bullfrogs (*Rana catesbeiana*) made their presence known in the spring with their deep mating calls. Without much effort Jacobs was able to show me that South Georgia was well represented in the amphibian collection by plenty of frogs, newts, and salamanders. But the real story here was that many amphibians are threatened by the activities of humans. Worldwide, some 120 species of frogs, toads, and salamanders have disappeared since 1980.

in Douglas. The county land records showed that the farm was more than eight thousand acres, and Gaskin's gift of it to his daughter Annie Belle had occurred in 1898.

On a cool spring morning I left Douglas with Chris and Frankie, traveling north to see landmarks chosen based on my expressed interest. Frankie had set the itinerary in the most efficient order, beginning with Broxton Rocks Preserve, a venerable site where we had picnicked as children with our parents. This site is the single largest exposure of a sandstone formation known as Altamaha grit, named for the Altamaha River. Here Rocky Creek, a black-water stream, wends its way through a series of rock outcroppings along a bed carved over thousands of years and plunges over a delightful waterfall in the middle of it all. It is the southernmost such feature in Georgia. Plants thrive in cracks of the rock because of water dripping down

Broxton Rocks waterfall, southernmost in the state of Georgia.

the rock faces. Many are rare and, in some cases, endangered. Frankie, a botanist by profession, has documented the species for posterity.

The rocks and the waterfall were enough of an attraction for families like mine to visit on a day trip, but in addition the place has a special feel. I can only compare it to my first trip into an ancient redwood forest. You instinctively know that something profound needs preserving. Fortunately, with the help of Frankie, the Nature Conservancy purchased the tract and is protecting it for the future.

Leaving Broxton Rocks Preserve, Frankie said we would drive to the site of my grandfather Clough's farm, which turned out to be not far away, just over the Coffee County line in Jeff Davis County. The county road we traveled as we left Broxton Rocks was paved until it encountered a crossroad. On the other side of that intersection, the road was dirt and consisted of two ruts made over the years by wagons and trucks. This, Frankie said, marked the southern boundary of the Clough farm. According to the plat maps Chris had brought, the northern boundary of the farm was the Ocmulgee River. The area between the road and the river was rural when my grandparents lived here, and it is rural today, with no residences nearby. The land is sandy and not suited for raising most crops, but in my grandfather's day it was populated by longleaf pine and cypress—prime species for lumber.

I asked about the signage at the entrances to the property and learned that the land is now protected by the Georgia Department of Natural Resources because it is home to two threatened species: the eastern indigo snake and the gopher tortoise. Frankie said work was underway to restore the longleaf pine to this land, where it once was the dominant species. All of this new information caught me by surprise. For the first time in my life, I was seeing the place where my grandparents and their family, including my father, had made their home before the Great Depression, and I was being told it was now protected because of a snake and a tortoise. It was a lot to process all at once.

Shortly after we arrived at the site of the farm, we were joined by Dirk Stevenson, a herpetologist and the director of the Orianne Society, which has as its goal the conservation of the indigo snake and its habitat. We asked Stevenson if there were any indigos nearby. He said yes and immediately walked off in search of gopher tortoise burrows, where indigos tend to live. We heard a shout and found that Stevenson had used his snake stick to fish out a beautiful indigo, which he said was a female. I had no idea how he came to this conclusion, but I took his word for it. Our girl was about

Herpetologist Dirk Stevenson (*left*) walks the old Clough farm site in Jeff Davis County with Frankie Snow and the author, March 2014.

five feet long, and she coiled up on Stevenson's arm, showing no signs of panic. Her eyes seemed to be studying us to take our measure and evaluate any need to fear us or put us in our place. You could see the reddish-brown coloring under her face and the lustrous indigo underside, which are trademarks of the species.

Stevenson suggested that I hold her and gave me some instructions, which I did not fully comprehend as I took possession. At first the snake coiled around my arm as she had with Stevenson. However, she sensed almost immediately that I was not a snake expert, loosed her coil, and dipped her head toward the ground. Stevenson said not to worry; she was unhappy about being out in the sun and was looking down at a burrow hole, where she longed to be. After allowing the snake to return to her

The author holds a female eastern indigo snake at the old Clough farm site, March 2014.

home, I asked Stevenson if he had ever been to the Smithsonian, and he said yes, a while back. Then he volunteered that recently keepers from the National Zoo had been to this area to collect a healthy pair of male and female indigo snakes to be mated as part of the zoo's breeding program for endangered species. This news hit me like a gunshot. Not very long ago on my visit to the reptile collections, curator Jeremy Jacobs had mentioned that the National Zoo might have live indigo snakes from South Georgia. Now I learned that they had come from the place where I was standing at that very moment, my grandfather's long-lost farm. Yet another remarkable coincidence on my journey of discovery.

When I got back to Washington, it did not take long to schedule a visit to the National Zoo to meet my fellow South Georgia natives. The keeper introduced me to the indigo snakes that had grown up in the sandy soils of my grandfather's farm and said they were healthy and would mate sometime next year. While everyone around me was thinking about the indigo snakes, I was thinking about their connection to a place where my family history runs deep.

## And Then There Was Rocky Hammock Landing

After seeing the indigo snake, we drove around the farm property to places where we could see the imprint of former dwellings: foundations overgrown with vegetation and barely standing chimneys. Ornamental shrubs, now looking wild, indicated where yards used to be. My father grew up here. He hunted in these woods and fished in the nearby Ocmulgee. I felt his presence.

Next, Frankie guided us down a dirt road that he said would take us to Rocky Hammock Landing, named for the outcroppings of Altamaha grit that protrude from beneath the banks of the Ocmulgee River, providing shallows where riverboats used to land. As we consulted the plat maps, it became clear that Rocky Hammock Landing was actually located along the northeast boundary of the old Clough farm. I realized that the two places I had heard about for years were in fact two pieces of a whole. When that riverboat carrying Grandfather Johnson and his family arrived at Rocky Hammock Landing in 1904, my mother had stepped off the boat onto the farm where my father lived. I had not only found the two sites I was looking for; I had found a singular place where my existence was preordained.

I watched the muddy Ocmulgee River sluggishly pass by Rocky Hammock Landing, and I thought about all I had learned that day. Why had my parents never brought me to see this place, even though it was only twenty miles from Douglas? We had passed fairly close to it many times visiting relatives on their farms. Why not go just a few miles farther to see where Grandfather Clough's farm had been and where Mom had arrived with her family at Rocky Hammock Landing? I don't know the answer. I can only imagine that it was painful for Dad to remember what life was like before his family's farm was lost, and he and his siblings were forced to leave the only home they had ever known. It was like a chapter of his life that was closed, and he had no need, or wish, to reopen it.

In the end, I take comfort that the Clough farm, once a symbol of sadness and loss, is now a protected natural area in the process of being returned to a place of renewal. My parents, who passed away years ago, are not around to learn the good news. Then again, they might somehow know that I finally visited the old farm.

## Tending the Essentials

Amphibians and reptiles are not cuddly sorts of animals, but they are essential players in our natural systems. Lacking the mobility of other species, they are vulnerable to changes in the environment and their habitat. The Smithsonian collections tell the stories of species that are endangered or threatened, including turtles, snakes, tortoises, and salamanders. Loss of habitat, overhunting, and climate change are all elements in these stories. In the past two hundred years we almost lost the alligator, a keystone reptile that helps support many other species.

Saving these animals may not seem important to some, but they are part of the interconnected natural world on our planet. Their stories also connect to the people, such as my own family, who possess a title to the land for a period of time. For the amphibians and reptiles, the land is everything. Without access to it, they cannot exist.

CHAPTER EIGHT

# So Many Specimens, So Little Time

## *Insects, Trees, Plants, Fishes, and Crawfish*

Key in the word "Okefenokee" on the Smithsonian Collections Search Center website and more than eight hundred listings pop up, ranging from natural history specimens to paintings. What is impressive about the listings is that most were added to the digital collections while I was writing this book. Sprinkled among the Okefenokee results are references to the comic strip "Pogo" by Walt Kelly, which entertained Americans for four decades. An amiable opossum (*Didelphis virginiana*), Pogo lived in the Okefenokee Swamp with friends Albert Alligator, Porky Pine, and Churchill "Churchy" LaFemme, a mud turtle. Pogo is well represented in the Smithsonian collections, not only in the Smithsonian Libraries, but also in the entomology collections, which contain a type of weevil found in South Georgia named after Pogo.

Pogo was something of a philosopher, and one of his best-remembered lines is "We are surrounded by insurmountable opportunities." I thought of those words as I considered the flora and fauna of the Smithsonian's natural history collections that remained for me to investigate. There were so many specimen types that an inclusive discussion threatened to swamp this book. I decided the only solution was to use a broad brush and include insects, trees, plants, fishes, and crawfish all in one chapter. (The latter are my nod to invertebrate zoology, which is about a lot more than crawfish, but my memories of crawfish are more appealing than my experiences with worms and mollusks.) This task proved easier said than done, because once you open its doors, each collection offers temptations to those who are attracted to opportunities to dig deeper and learn more.

## Insects

Insects are the Jekyll and Hyde characters of nature. When a mosquito buzzes around us, scouting out a vulnerable spot from which to drink our blood, we think, "This is one species that Earth's web of life could do without." However, when a colorful butterfly flits by in its whimsical flight, we are warmed by thoughts of nature's grace and beauty. Insect behaviors often seem bizarre, such as when a female praying mantis eats the male who has just mated with her. Such behavior may cause unpleasant nightmares for the males among us, but it is just one of many strange adaptations of the insect world. As Annie Dillard wrote in her book *Pilgrim at Tinker Creek*, "Fish gotta swim and birds gotta fly; insects, it seems, gotta do one horrible thing after another."

Nevertheless, we still experience pleasure watching a bumblebee (*Bombus*) flying purposefully as it collects nectar, spreading pollen from flower to flower and making it possible for plants to produce another generation. As a boy I was fascinated by bumblebees, whose wings beat 250 times a second, allowing them seemingly to float through the air. Few insects have inspired music, but I grew up hearing my sister play Nikolai Rimsky-Korsakov's "Flight of the Bumblebee," and it never failed to delight.

As I worked on this book, I made sure to see the Smithsonian collection of bumblebees, some sixty-six thousand strong. It was one of the first natural history collections chosen for a major digitization effort because these creatures are critical as pollinators. Unfortunately, they are at risk today because of the indiscriminate use of pesticides.

Hundreds of other insect species are represented both in South Georgia and in the Smithsonian collections, and many could be singled out for this chapter. Gnats (*Liohippelates bishoppi*), ever present outdoors in South Georgia summers, were considered, but I already gave them their due when discussing the gnat line (chapter 2). I would only add here that South Georgians do have a sense of humor about the annoying creatures with which they are obligated to live. For example, the good citizens of Camilla, Georgia, honor gnats with their Gnat Days Festival. This popular event joins other South Georgia festivals featuring wild hogs (Abbeville), fire ants (Ashburn), swine (Bainbridge), rattlesnakes (Whigham), and mules (Calvary). Collectively they provide entertainment opportunities lacking in big cities.

Choosing which insects to highlight from the collections was like picking out a few drops to represent water, since the entomological collections number thirty-four million specimens. In the face of such vast numbers, it was impossible to be inclusive, so I chose two for personal reasons: yellow jackets (Vespidae) and wingless "ants" (Mutillidae). Both are wasps by another name. The yellow jacket was chosen out of loyalty to my undergraduate alma mater, the Georgia Institute of Technology, and the wingless ant fascinated me as a boy.

While most people really don't care for either of these insects, there are those who love them. They are called Smithsonian curators, and they work in the Entomology Department of the National Museum of Natural History. I owe a debt to Torsten Dikow, Sean Brady, and Floyd Shockley, who took the time to find yellow jackets and wingless ants that fit my bill.

### *The Yellow Jacket: Beloved Mascot or Irritating Insect?*

Georgia Tech has two sometimes competing mascots: an insect and a car. The insect is a yellow jacket; the car is a 1930 Ford Model A sport coupe, otherwise known as the Ramblin' Wreck. The Ramblin' Wreck derives its status from the famous fight song line "I'm a ramblin' wreck from Georgia Tech and a hell of an engineer . . ." The story behind the yellow jacket is more complicated.

The yellow jacket moniker did not originate with the insect. It came about in the early 1900s when a group of Georgia Tech boosters wore yellow coats to athletic events, demonstrating their support for the team, whose colors were gold and white. Sports reporters took to referring to Georgia Tech fans as "the yellow jackets." Since coats are not mascot material, the insect was eventually designated for this role.

For many years the Georgia Tech yellow jacket was primarily depicted in drawings and sketches. But in the early 1970s universities began to adopt mascots that paraded around at ball games and mugged for television cameras. It was inevitable that Georgia Tech would finally acknowledge its indebtedness to the genus *Vespula*. In 1980, Buzz debuted in the form of a yellow jacket costume worn by a student who was good at keeping fans' spirits up even when their team was losing. Buzz has developed a lively personality and has won national awards.

The insects popularly called yellow jackets are predatory wasps. Several different yellow jackets are native to South Georgia, primarily the eastern (*Vespula maculifrons*) and southern (*Vespula squamosa*). A menacing, invasive wasp is also making headway into the region, the German yellow jacket (*Vespula germanica*). One way to distinguish among these species is through the coloration and appearance of body patterns. I may be biased,

Southern yellow jackets from South Georgia. National Museum of Natural History entomology collections.

but I have to say the southern yellow jacket is the best looking and most colorful.

The Smithsonian collections contain more than eleven thousand yellow jacket specimens, some of which, as I point out to my fellow Georgia Tech alumni, come from Atlanta. There are also many from South Georgia, mostly of the southern type. It was eerie to see hundreds of the little fellows arrayed in boxes in neat lines, each held in place by a pin through its thorax. Each specimen was identified by the place where it was collected and by the person who collected it.

Regardless of their family, all yellow jackets are aggressive, particularly in defending their nests. The nests can be in the ground, in rotting logs, or hanging on the walls of houses or barns. Females are not only the queens but also the workers, while the males have one job, mating. If you are stung by a yellow jacket, a female did the deed, and because the female's stinger is not barbed she has the ability to sting multiple times. The yellow jacket's

gardens were home to ten more tree species that grow in South Georgia, including American holly (*Ilex opaca*), black gum (*Nyssa sylvatica*), laurel oak (*Quercus laurifolia*), loblolly pine (*Pinus taeda*), red maple (*Acer rubrum*), southern red oak (*Quercus falcata*), sugarberry (*Celtis laevigata*), sweetgum (*Liquidambar styraciflua*), sycamore (*Platanus occidentalis*), and tulip poplar (*Liriodendron tulipifera*). She went on to note that if she had more time, she could probably come up with a few more. I was sure she could, but I was thoroughly impressed with her list as it stood.

### *The Botany Collections*

While the longleaf pine was not among the living collection of the Smithsonian Gardens, I had hope that I would find a specimen of it in the botany collections. As my travels through life took me farther from my childhood home, one memory remained as strong as the day it was formed—that of the longleaf pine. Graceful giants that whispered in the breezes, they seemed to speak for the ancients, to hold knowledge of the past that hovered just beyond our reach.

Evolved over millions of years as a fire-resistant species, the longleaf pine was logged or cleared for agriculture or lost its habitat to development. Today it occupies less than 2 percent of its original range. Fortunately, the tide is turning, and many groups are working to restore enough of the trees to enable them to survive into the future.

When I visited the herbarium, collections manager Rusty Russell immediately made the trip worthwhile. He said that not only did the collections include a longleaf specimen from South Georgia, but it had been collected in 1903 in my hometown of Douglas! The specimen had been cropped from the end of a pine limb. Its fifty or so needles were now brown with age, but they were arranged elegantly in groups, some with many needles and some with only a few. Alongside the needles lay a small pinecone, well shy of the size it would have reached at maturity. The arrangement was expressive and at the same time spare, like a Zen painting. The name of the collector was Roland Harper who, it turns out, is famous as one of the most important botanical collectors in the South. In 1903 when he was in Douglas, Georgia, he was a young man working on his PhD at Columbia University.

Harper was the son of German immigrants who eventually established a home in Georgia, in part to improve Roland's health because he had been diagnosed with tuberculosis. Harper graduated from the University of Georgia in 1897 as an engineer, but it did not take him long to choose botany over engineering as a career. In 1900 he received a scholarship from

Longleaf pine specimen collected in Douglas, Georgia, by Roland Harper, 1903. National Museum of Natural History botany collections.

Columbia University and set off for South Georgia to document the flora of the region defined by the location of the Altamaha grit rock formation. His work was grueling, particularly in the summer, but he was meticulous in his documentation. Along with making sketches, he took several thousand photographs, some including scenes from Douglas and its surroundings. One picture is of the area around Gaskin Springs Holiness Campground, named for one of my relatives on my great-grandfather Daniel Gaskin's side of the family.

A copy of Harper's dissertation, "Phytogeographical Sketch of the Altamaha Grit Region of the Georgia Coastal Plain," is in the Smithsonian Libraries natural history collection. The libraries also have a handwritten manuscript by Harper about an exploratory trip he made to the Okefenokee Swamp, which has one of the longest titles I have ever seen: "Journal of a Botanical Exploration of the Okifinokee [*sic*] Swamp and Adjacent Territory in the Counties of Ware, Pierce, Charlton, Wayne, Glynn, Camden, Clinch, Echols and Lowndes in Georgia and Portions of Nassau, and Baker during the Month of August 1902"—and it continues.

Harper went on to become known for his prolific work on species he observed in South Georgia, Alabama, and other southern locales. He gave hundreds of his specimens to the Smithsonian, where they stand out because of the clarity of his presentation. In addition to Harper's longleaf pine specimen from Douglas, the collections contain fifteen other specimens from Coffee County. All display his consistent attention to detail and aesthetics.

Harper was employed by the Geological Survey of Alabama until his death. An interesting aside is that he waited until he was sixty-five to get married. If Harper had stuck with his interests in botany, his reputation would have been untarnished, but later in life he elected to speak out on subjects such as eugenics and communism, which laid bare his prejudices and biases for all to see. Flawed as he was for his personal convictions on other issues, his work on botany stands as among the best of his time.

Beyond Harper's specimens, the botany collections yielded specimens of two other plants I was hoping to find: wiregrass (*Aristida stricta*) and rabbit tobacco (*Pseudognaphalium obtusifolium*). I learned that Native Americans and early European settlers used rabbit tobacco as a medicinal herb; it was believed to cure a range of illnesses. Native Americans also used rabbit tobacco smoke to identify ghosts and bad spirits, so they could be forced to leave a dwelling.

Altamaha grit rock face with unusual plant species nourished by dripping water. Frankie Snow (*left*), Jim Cottingham (*right*), and the author at Broxton Rocks.

Roland Harper was one of the first botanists to visit Broxton Rocks, the large outcropping of Altamaha grit where my family had picnicked when I was a child. There, fissures in the rocks by the stream provide cool, moist crevices where unusual plants grow. While he knew the place supported a number of rare plant species, there is no indication that Harper took the time to document them. The South Georgia botanist Frankie Snow did this task more recently. Of particular interest to me was the green-fly orchid (*Epidendrum magnoliae*). The only epiphyte in the orchid family that exists in the wild northern part of Florida, it was identified by William Bartram as the "tree orchid."

Given that the Smithsonian has an orchid collection with more than eight thousand species, I thought the green-fly would be included. To my surprise, it was not. I put Frankie in touch with Dennis Whigham, a scientist who heads a new Smithsonian initiative to save endangered orchid species that are native to the United States, and the two agreed on a protocol for the donation of a green-fly orchid. After checking with folks who guard endangered species to make sure the green-fly could be collected, Frankie sent Whigham the reproductive parts of the orchid, so he can grow it for the Smithsonian. The green-fly was Frankie's second donation to the Smithsonian collections that resulted from this book. His first was specimens of Altamaha grit for the rock collections.

## Fishes and Crawfish

Given the time I spent as a boy in swamps, ponds, creeks, and rivers, I was well acquainted with fishes and crawfish. But my knowledge of freshwater fishes was mainly about those we caught to eat, and I only knew that crawfish were experts at building mud chimneys around their holes because they popped up so many places where I wandered. Fortunately for my edification, the Smithsonian had documented the species of South Georgia through a series of collecting expeditions and with the help of volunteers. Georgia provides fertile ground for ichthyologists because it has one of the most diverse stocks of fish species in the United States. The collection manager of the Division of Fishes, Jeff Williams, was there to help and provided lists of fish species in the collections.

Not to be left out are the crawfish, which are well represented in the crustacean collections. I relied on the work of former Smithsonian scientist Horton H. Hobbs. He is deceased, but his book *The Crayfishes of Georgia* gave me all I needed to know.

### *Fishes*

Fishes have been gathered from South Georgia for the Smithsonian collections since the 1880s. I wish I could say the collectors made it to my favorite boyhood haunt, Shop Pond, but as far as I can tell they did not. However, specimens were collected from Seventeen Mile Creek, which rises not far from Douglas in Coffee County.

I am going to violate the scientific rules and divide the fishes into two groups: small ones that are used mostly for bait or aquariums and those big enough to eat. Until Jeff Williams gave me his long list of specimens

Okefenokee pygmy sunfish.

from South Georgia in the collections, I had no idea that the small fishes we called minnows and watched darting around in the shallows are in many cases prized for aquariums. They include the beautiful Okefenokee pygmy sunfish (*Elassoma okefenokee*), its cousin the Everglades pygmy sunfish (*Elassoma evergladei*), the blue-spotted sunfish (*Enneacanthus gloriosus*), its cousin the banded sunfish (*Enneacanthus obesus*), the golden shiner (*Notemigonus crysoleucas*), the pygmy killifish (*Leptolucania ommata*), the lined topminnow (*Fundulus lineolatus*), the eastern mosquitofish (*Gambusia holbrooki*), and the swamp darter (*Etheostoma fusiforme barratti*). All of these wonderful small fishes are a testament to the bounty of nature in South Georgia.

Most of the bigger specimens from South Georgia in the Smithsonian collections are fishes I can remember catching. They include an array of sunfish, such as the mud sunfish (*Acantharchus pomotis*), dollar sunfish (*Lepomis marginatus*), green sunfish (*Lepomis cyanellus*), warmouth (*Lepomis gulosus*), and flier (*Centrarchus macropterus*). Then comes the lake chubsucker (*Erimyzon sucetta*), the black crappie (*Pomoxis nigromaculatus*), and the top-of-the-line largemouth bass (*Micropterus salmoides*). All of these fishes make a great meal, especially when you are camping and have been out all day working up an appetite.

Of course, just about all the different catfish, little and big, are found in South Georgia and are represented in the collections. The madtoms (margined, *Noturus insignis*; speckled, *Noturus leptacanthus*; and tadpole, *Noturus gyrinus*) are small guys, and some are sought by aquarium keepers for their habits, not for their grace or beauty. Madtoms, like their bigger counterparts, will eat most anything and like to hang around the bottom of aquariums or in small caves or nooks. They are interesting and provide an aquarium with a throwback look.

The big catfish from South Georgia in the collections include the black, yellow, and white bullheads (*Ameiurus melas*, *Ameiurus natalis*, and *Ameiurus catus*, respectively) and the channel catfish (*Ictalurus punctatus*). They all share the same trait of eating whatever reaches the bottom of the water body where they live. If they come from relatively clean water, they cook up into a great meal, especially with hush puppies. Catfish even have a festival held in their honor in Kingsland, Georgia, near the coast by the Florida border. Appropriately named the Kingsland Catfish Festival, it has been going strong since 1983.

Fishes in Georgia are doing well overall. However, the numbers of some species are in decline, principally due to loss of habitat, pollution, and siltation from agricultural lands and construction sites. Climate change will become a problem for some species as the sea level rises and saltwater intrusion threatens their habitat. Increases in water temperature and acidity, also from climate change, may cause the demise of species, like trout, that are sensitive to environmental changes.

### *Crawfish*

Crawfish are also known as crayfish and crawdads, and they are called mudbugs for their habit of building mud chimneys around the holes to their underground burrows. A hardy group, they eat just about anything, including dead fish. Crawfish are curious creatures and like variety in their environment, including finding a spot where they can breathe air from time to time. They come in different colors, including red, brown with spots, blue, and green. They even change colors depending on the season of the year.

Scientifically, crawfish in Georgia belong to the genus *Procambarus*. Before 1942, *Procambarus* was a subgenus, but Dr. Horton H. Hobbs Jr., who was passionate about crawfish, documented so many species that they were moved up to genus status with the big boys. In his lifetime Hobbs named 168 crawfish species and became known as "Old Man Crawfish" or "Crawdaddy."

Hobbs was born in Florida and was interested in nature as a child. He earned his doctorate from the University of Florida and taught for a time at the University of Virginia. Eventually his career took him to the Smithsonian, where in 1959 he was appointed senior scientist in the Department of Invertebrate Zoology in the National Museum of Natural History. He held this position until he retired in 1984. Hobbs spent a lot of time in Georgia and wrote the seminal publication, *The Crayfishes of Georgia*, in which he noted that the state was home to sixty-six species. From the 1940s through the 1970s, he collected extensively in South

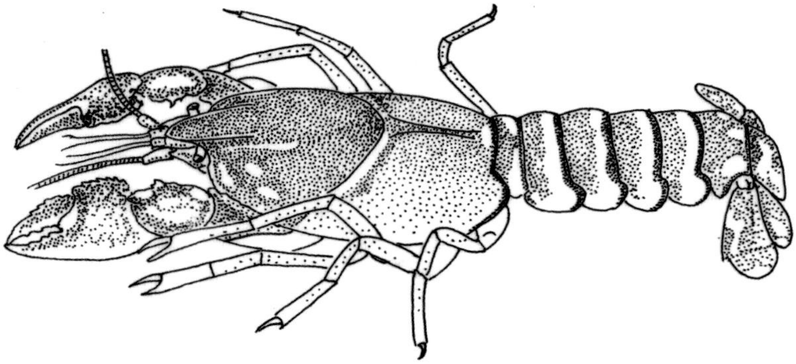

Christmas tree crayfish collected in South Georgia by Horton Hobbs. Sketch by Hobbs. National Museum of Natural History invertebrate zoology collections.

Georgia, including in and near Coffee County. He always had his collection equipment with him in the trunk of his car. His field notes cited typical collection spots, such as creeks and small pools in swamps, but just as often the site was a roadside trench or a drainage ditch. I have not counted them all, but his specimens from South Georgia number in the hundreds, including more than a few holotypes that identified new species. Among these is *Procambarus pygmaeus*, which Hobbs named in 1942, a charming little fellow known to aquarium collectors as the Christmas tree crayfish because it sports a green color tinged at the joints by a delicate red hue.

When I was growing up, folks in South Georgia did not much cotton to eating crawfish. I learned about eating them when I first visited my girlfriend (and wife-to-be), Anne Robinson, at her home in New Orleans, where she had moved with her mother after her father died. Cajun cooking in and around New Orleans produces a range of tasty crawfish dishes, including po' boy sandwiches, pies, crawfish boils, or, if you want fancy, étouffées.

To move South Georgians toward more crawfish in their diet, the town of Woodbine, located along the coast a little south of Brunswick, hosts the Woodbine Crawfish Festival each year. It is steadily growing in popularity along with the crawfish dishes. I may give it a try soon.

## Many Species, One Natural World

This chapter has covered a lot of territory. Although it deals with living things, they do not necessarily have a lot in common. Insects, fishes, and crawfish are all animals, but they diverge pretty quickly after that. The same is true for trees and plants. What brings them all together is that they share the natural world with humans, and we need all of them as part of the life cycle of that world.

With 91,000 species in North America, insects are the most diverse group of this chapter and cumulatively make up the largest biomass on Earth. They also make up the largest set of collections at the Smithsonian, 35 million specimens and counting. Because insects tend to like hot, humid, semitropical conditions, South Georgia is well represented in the collections. The botany collections are 5 million strong, and the fish collections include 4 million specimens. Crawfish are part of the invertebrate zoology collections, which number more than 1.7 million specimens. In an odd way, bunching all of these collections into one chapter shows both the incredible range of the Smithsonian natural history collections and the wonderful diversity of the natural history that makes up my home.

Finally, this chapter again reminds me of the remarkable people, like Roland Harper and Horton Hobbs, whose passion for science and specimens provided us with collections that allow us to appreciate and understand our world. I was glad to see Frankie Snow added to this list when he donated a green-fly orchid from Broxton Rocks to the Smithsonian.

CHAPTER NINE

# Geology

## *Rocks, Minerals, Meteorites, and the People Who Love Them*

South Georgia has no mountains, earthquakes, or deposits of gold or silver, so geologists, rock hounds, and fortune hunters have pretty much passed it over in favor of more profitable climes. If this is not enough, even though thousands of meteorites fall to earth each year, the official meteorite listings for the United States cite only nine with coordinates that put them in South Georgia. As with rocks and minerals, the official meteorite cupboard is pretty bare for my home. However, being a geological engineer and a glass-half-full kind of guy, I could not pass up a search for South Georgia connections in the geology and meteorite collections of the Smithsonian. Besides, I had learned that mining the Smithsonian collections could yield treasure, even when it was not obvious that it was there.

## Minerals and Rocks

One of the early roles of the Smithsonian was to identify deposits of ores and minerals with economic value. This responsibility continues to this day, resulting in one of the world's most comprehensive collections of these materials. When it comes to rocks and minerals at the Smithsonian, the obvious place to begin is with Leslie Hale, the collections manager for the National Rock and Ore Collections. Like so many other colleagues, Hale and her fellow Smithsonian scientist Ben Andrews were ready and willing to help in my search.

### *South Georgia Minerals*

With the help of Hale and Andrews, I learned that three important geological commodities are mined commercially in South Georgia. I knew about one of these, the clay mineral kaolinite, from my studies as a geological engineer. I had little experience with the other two: bauxite and rutile (a mineral composed primarily of titanium dioxide). The collections include specimens of all three of these materials, obtained many years ago.

It turns out that if you are interested in mining kaolinite and bauxite in Georgia, a good place to look is just below the fall line, noted earlier as the northern boundary of South Georgia. Kaolinite and bauxite are products of the weathering of rocks found in North Georgia. Streams and rivers flowing down the slopes of North Georgia carry the minerals south in their sediment load. At the fall line, where the land flattens and the water slows, the sediment and its valuable contents settle out. The kaolinite deposits are large enough to be significant to the world market, but the bauxite deposits only support a small level of production.

Clay minerals such as kaolinite form sheet-like structures that give them characteristics not found in other minerals, and South Georgia is home to some of the richest kaolinite deposits in the world. The ultrawhite color and texture of the mineral make it ideal for use in producing fine china and porcelain. Discovered in the mid-1700s, Georgia kaolinite quickly became an important commercial export for the new colony, and it was sought by manufacturers, including Josiah Wedgwood, who used it for his newly popular china and porcelain products.

Beyond its use in making china and porcelain, kaolinite is used to make the coating on glossy white paper and is even in items you find in your medicine cabinet—toothpaste, cosmetics, and remedies for upset stomachs. Each day most Americans ingest a little of the soil of South Georgia when they brush their teeth. Welcome to my home!

The other mineral from South Georgia in the Smithsonian collections that is worthy of mention is rutile. Rutile is found along the coast of Georgia where the coastal plain deposits are thick, and the titanium dioxide it contains is important because of its texture and color, often called "perfect white." As a pigment, it is used in paints, sunscreens, food colorings, and plastics, all of which we come in contact with daily.

Although I did not know it until recently, titanium dioxide was used for years in small amounts to whiten the wonderfully sweet covering on powdered sugar donuts. However, Dunkin' Donuts, for one, has announced it will no longer use titanium dioxide out of concern that ingested particles might find their way into the cells of our bodies. As with kaolinite, Americans who've been eating powdered sugar donuts through the years carry a little bit of South Georgia in their bodies.

A discussion of titanium dioxide brings us to the Okefenokee Swamp. The Okefenokee has long served as a natural resource for humans, beginning with the first migrants who came to the New World thousands of years ago. As cited in chapter 8, it is home to a remarkable ecosystem with species found in few other places. With the arrival of Europeans, large-scale logging reached into the Okefenokee and took its toll, but fortunately it was protected in 1937 as a national wildlife refuge.

In the 1980s E. I. du Pont de Nemours and Company discovered a deposit of titanium dioxide just to the east of the Okefenokee preserve and began planning for its extraction. Although assurances were given that environmental concerns would receive full consideration, conservation groups came together to ask critical questions. Because so little is really understood about how the swamp's natural water systems work, doubts were raised about whether any form of protection would be adequate for its fragile ecosystem. The federal government also weighed in to protest the project. In 2003, after years of debate, DuPont decided to set aside its plans and, in a provident move, "donated" the 16,000 acres of land it had acquired as an addition to the 396,000-acre Okefenokee National Wildlife Refuge. A contingency to the "donation" requires ninety million dollars to be raised from federal, state, and private sources over time to compensate DuPont.

The discussion that led DuPont to decide not to implement its mining plans near the Okefenokee Swamp was a positive step for the emerging bioconservation ethic in South Georgia. However, this case is unlikely to be the last effort to access the natural resources of the region. Minerals like titanium dioxide have a way of finding new uses over time.

### *South Georgia Rocks*

When it comes to rocks in South Georgia, Altamaha grit is high on my list, not only because it has little competition on the coastal plain, but also because it touches the story of my life in two places: Broxton Rocks in Coffee County, where my family enjoyed day outings, and Rocky Hammock Landing along the Ocmulgee River on what was once my grandparents' farm. The "grit" is actually sandstone that formed some twenty million to thirty million years ago. As rising and falling seas laid down layer upon layer of sand and sediment, some of it was compressed into Altamaha grit. It ranges from hard and flinty to crumbly and sandy. Because the rock itself has little commercial value, it remains on the fringes of interest in geological circles.

Perhaps because of its relative obscurity, Altamaha grit attained a unique status in my quest. It was the first object I sought in the Smithsonian

Back in Enigma, the town was beginning to awaken, and a few folks were appearing in the stores. We drove into the parking lot of the Enigma Café where the sign advertised "American and Mexican food, Monday to Saturday, 10 to 3." Rowan owns the café, but he closed it two years ago for lack of business. Not letting a resource go to waste, he uses it as an office. As we entered the café, we found not only Rowan but also an enterprising young reporter from the *Tifton Gazette*, Latasha Ford.

Bobby Rowan welcomed us, and we talked a while about old times when he was still in the state senate and I had just arrived as president of Georgia Tech. He retired from the senate in my first year but returned often to assist local agricultural interests. It was always a pleasure to see him. Rowan was a member of a large family with nine children. He said about his father, "He was a better mule farmer than a tractor farmer," meaning he was more at home with traditional approaches than with new technology.

I showed him where we had been that morning and told the story of the meteorite. He said he had never heard about the meteorite, but his second cousin owned the land where it was found. We both thought it odd that so little was known about the meteorite in such a small community, but it seems the fate of the Enigma meteorite to be a bit of an enigma even in its hometown. Our conversation left Rowan feeling proud that his hometown had its own meteorite in the Smithsonian's permanent collections.

I had brought a gift for Rowan for meeting with us, but it was a Georgia Tech hat, and I realized he was a graduate of the archrival University of Georgia. Gentleman that he is, he said he was glad to get it, since one of his brothers had graduated from Georgia Tech. As he put it on, he added that his brother "was the smart one."

Following my visit, the Enigma meteorite was finally mentioned in the newspapers. An article written by Latasha Ford was published in the *Tifton Gazette* under the headline "Men Meet to Discuss Enigma Meteorite," and a copycat article ran later in the *Atlanta Journal-Constitution*. The heretofore ignored Enigma meteorite finally had its day in the sun.

### *The Pitts Meteorite, a Flashy Cousin*

Pitts, with a population of about four hundred, is located in Wilcox County. It thrived initially after its founding in 1888 because the Americus, Preston and Lumpkin Railroad had been encouraged to pass through, and workers made their homes there. The name of the town was originally Kings Crossing, an elegant choice at the time. But a parsimonious postmaster general complained that it was too long, and it was changed to Pitts to honor Ashley J. Pitts, a merchant in the area.

Around nine o'clock on the morning of Wednesday, April 20, 1921, Pitts became ground zero for an encounter with a meteorite that arrived in style: a bright fireball with a glowing tail was seen for miles around, accompanied by the sounds of explosions and a trail of smoke. The *Cordele Dispatch and Daily Sentinel* covered the event with a special edition that appeared on the afternoon of April 20:

> GREAT METEOR ON VISIT TO EARTH AT EARLY HOUR, PEOPLE OF COUNTRYSIDE WERE GREATLY EXCITED
>
> One of the most exciting incidents connected with the life of this generation was the call of a tramp meteor near nine o'clock this morning. It came from the east and seemed to be traveling like the deep echoes of a heavy thunderstorm and burned a streak of fire on a jagged course into space.

Fortunately, S. W. McCallie, the state geologist of Georgia, was made aware of the meteorite early on. He received a piece for examination from a local citizen on April 22, and by April 24 he was in Pitts reconnoitering the area and meeting with eyewitnesses. He wrote a detailed description of the events and his investigation, which was published in the *American Journal of Science and Arts* in 1922. As he put it:

> The fall of the Pitts meteorite here described was the most striking phenomenon of its kind heretofore observed in the State of Georgia. . . . It was no doubt plainly visible over an area of several thousand square miles, and could have been distinctly seen by fully a quarter of a million people had they been looking in the proper direction.

As to eyewitness reports, he wrote:

> The dense smoke in the wake of the flaming fire ball was referred to by the Albany and Moultrie witnesses as a luminous trail following the flaming ball.
>
> In the immediate vicinity of Pitts the sound was described as several loud explosions, causing the earth to tremble, followed in quick succession by a number of lesser explosions.
>
> This specimen [referring to the third of four pieces] fell within three feet of a boy walking along the road, who was covered with earth.

McCallie confirmed that the meteorite fell in four pieces, and he provided a map that gave the relative locations of each piece based on distances to landmarks, such as farmhouses. For example: "The largest piece weighing 57 ounces fell within less than 75 yards of Nancy Brinson's house where it was dug up a few minutes later still warm." McCallie concluded, despite some contradictory statements by eyewitnesses, that the meteorite was traveling in a northeasterly direction.

The meteorite's composition was analyzed by a chemist working for the state of Georgia, and a piece was sent to the Smithsonian for examination. (Notably, this piece did not remain at the Smithsonian but was returned.) The assessments confirmed that the meteorite was 90 percent metallic with a specific gravity of 7.33. Because of the notoriety of the Pitts meteorite, experts have studied it in depth over the years. It is technically classified as an iron, fine octahedrite, with silicate inclusions (IAB).

Most of the pieces of the Pitts meteorite ended up in private hands, but the Smithsonian had an ace in the hole in S. H. Perry who in the 1920s became friends with George P. Merrill, the head curator for geology. Perry, born in Michigan in 1874, had a strong interest in science when he was a student at the University of Michigan. However, after his undergraduate studies, he earned a law degree and developed a successful career as an editor and publisher of newspapers, eventually becoming vice president of the Associated Press. Aside from his family, his greatest love was meteorites, which he collected and analyzed, even writing papers for professional journals. To his enduring credit, Perry was committed to helping the Smithsonian build its collections for the good of science. In 1939 he was able to acquire the second fragment (as identified by McCallie) of the Pitts meteorite, and he gave it to the Smithsonian in 1941.

The Pitts meteorite, an iron, fine octahedrite with silicate inclusions (IAB), made impact in Pitts, Georgia, on April 20, 1921. National Museum of Natural History meteorite collections.

On seeing the meteorite during my visit to the collections, I was impressed. The Smithsonian's piece is about palm-sized, and it is heavy for its size—so heavy that when it was placed in my hand, my hand automatically dropped downward. The surface is smooth and black, and where it has been cut to expose the interior, I could see a shiny metallic-banded matrix with several amorphous inclusions, as if an artist had covered it with cross-hatching. This banding is known as a Widmanstätten pattern, its striking mineral highlights aligned in tightly linked sets of parallel straight lines that intersect each other. The pattern is caused by the formation of crystals during the slow cooling of the meteorite's interior. All in all, the Pitts meteorite is a star.

### *Visiting Pitts, Georgia—Twice*

Armed with the information I had gathered about the meteorite and having seen a fragment of it firsthand, I set out on a pilgrimage to Pitts, joined

by Johnny Gibbons and Frankie Snow. The task seemed simple. Drive to the vicinity of Pitts, enter the latitude and longitude from the literature on the meteorite into my iPhone, mark it with a pin, and devise a route to get there. For the record, the latitude and longitude officially cited for the Pitts meteorite is 31° 57' N, 83° 31' W.

Pitts is a small town at the crossroads of State Route 215 and U.S. Highway 280, surrounded by moderately rolling hills dotted with agricultural fields. True to its railroad roots, the Heart of Georgia Railroad still runs east-west through town. We drove in from the east on Highway 280, and the first visual indication of the town was a large cemetery lying to the north of the road. This landmark guided us to Old Cordele Road, a dirt road we followed to the designated meteorite landing site. The route was easy to follow and uncomplicated, and we passed only one small brick farmhouse on the way. The site we sought seemed to be in the corner of an agricultural field. Eyes on the pin on my iPhone, we walked down a packed grass road and

Downtown Pitts, Georgia, looking north on State Route 215, February 2016.

emerged in a field. According to the pin on the iPhone, we had arrived at the location of the meteorite landing. Not a building or a farmhouse was in sight. We checked and rechecked that we were at the right place and took pictures from different angles to document our visit. I felt gratified that I had visited the very site where the fragment of the Pitts meteorite now in the Smithsonian collections had come to earth in South Georgia.

In retrospect, it would have been reasonable to question why one single latitude and longitude reading was considered adequate to define the landing site for a meteorite that actually had fallen as four fragments, but that thought did not occur to us at the time. Once home, I wrote up the visit, content that the fieldwork about meteorites was complete.

About two months later I got a call from Frankie Snow, who had some news. He had participated in a conference at South Georgia State College and met a gentleman named Mike Miller, a pharmacist and the owner of Rochelle Drugs in Rochelle, Georgia. Rochelle is located about five miles from Pitts, and the folks in Pitts, with no drugstore of their own, use Rochelle Drugs. As a result, Miller knows about everybody in Pitts. Miller turned out to be one of those people in a small town who has broad-ranging interests. In the case of the Pitts meteorite, he had a personal stake because his father had given him a small piece of it. When Frankie told Miller that we had found the site of the meteorite using the latitude and longitude reported in the scientific literature, Miller said he had heard that information was wrong.

Now this was a problem. If the latitude and longitude provided in the literature were wrong, then I had lots of pictures of me standing in a field with no connection to the Pitts meteorite. I was disappointed at first, but then grateful that I had learned about this problem sooner rather than later. I was in for a return visit to one of my favorite towns, Pitts. But how would I find the true site? Mike Miller said he knew three people who had personal knowledge of the meteorite, including one man who was 102 years old and as a boy had seen it fly over his house.

Fortunately, I had agreed to a speaking engagement in Douglas and was able to line up a side trip to Pitts around that date. So, on February 23, 2016, Frankie Snow and I took another car trip to Pitts, first swinging by Rochelle to pick up Mike Miller at his pharmacy.

### *First Stop: Mac Pearce and the Brinson Home*

Mac Pearce lives on a farm to the north of Pitts alongside State Route 215. Pearce's farmhouse greeted us with a cropped grass lawn and a tidy, well-kept appearance. At eighty-two years of age, he met us with a strong

Nancy Brinson house site, State Route 215, Pitts, Georgia, February 2016. This was where the largest fragment of the Pitts meteorite fell. Mac Pearce is with the author.

handshake and a friendly greeting. He explained that he bought the farm in the 1960s, raised his family there, and farmed for a living before retiring and having to sell out to a large combine. Pearce knew of our interest in the meteorite and took us to a clearing just up from his house. It was set back from the road and surrounded by four mature pecan trees.

This site, he said, was where the home of Nancy Brinson and her family had been located. McCallie had visited this place in 1921 and documented that the largest piece of the meteorite had fallen in the front yard of the house. When Pearce bought his land in the 1960s, it included the old Brinson home, but the last Brinson had already moved out of the area some time before. The house had been rented out, but later it was consumed in an accidental fire.

A short walk to the north of the homesite was Snow Hill Cemetery in a grove of trees. The Brinson name was prominent on several headstones, and Pearce said that members of the family came from Florida every year to tidy up the place and take care of the headstones.

His evidence, combined with McCallie's report, made it clear we had now visited the correct location for the landing of the first and largest fragment of the Pitts meteorite. It was about two miles west of the location

Snow Hill Baptist Church, State Route 215, Pitts, Georgia, the impact site of the second fragment of the Pitts meteorite. Frankie Snow is with the author.

defined by the official latitude and longitude in the literature about the meteorite. We thanked Mac Pearce for his help and set out to find the location of the second fragment, the one that is in the Smithsonian collections.

From the site of the Brinson home it was possible to use the map provided by McCallie to find where the second fragment had landed, since the Brinson site was the anchor for his descriptions of the other places. We drove back up State Route 215 toward town to a location about seven hundred feet from the Brinson house, which, according to McCallie, was where the second fragment had landed.

This site was on the east side of the road where several abandoned structures stood, among them what had formerly been the Snow Hill Baptist Church and an affiliated school. Historical records show that the congregation was founded in 1894 by the local African American population, who built the church over time. As we approached the site, it was apparent that the wooden school structure was on its last legs, while the church, made of red brick, was still standing. The church at one time was a well-built, handsome structure, but left to the elements it was deteriorating. We peered through broken windows, and beams of sunlight streaming from the opposite windows illuminated the sturdy pews, reminding us that people once made this a house of worship. Before we left, Frankie, Mike Miller, and I took pictures of what all the evidence suggests is the real location of the landing of the second fragment of the Pitts meteorite.

### *Second Stop: The Home of Ruth and Dock Cantrell*

Ruth and Dock Cantrell are long-time residents of Pitts. Miller had told us that Ruth's mother saw the meteorite flying over her home and that Ruth could tell us about her experience. Our directions took us to a surprisingly familiar location—the small, tidy brick farmhouse on Old Cordele Road we had passed the year before on our way to what turned out to be the wrong landing site for the meteorite. Ruth and Dock had been married seventy-three years, and during this time they had pursued careers to pay the bills. But they also spread the good word through their gospel music, with Dock playing guitar and Ruth playing the piano and singing alto. Ruth had recordings of their work, which she kindly shared with me.

I asked Ruth Cantrell about the meteorite, but before she recounted her mother's eyewitness report, she surprised me by showing me some photographs from the Smithsonian meteorite collections. There it was in the photographs—the same meteorite I had held on my visit to the collections. Cantrell had cowritten and edited a comprehensive history of Pitts. Although mostly about the people of the area, it included a section on the arrival of the meteorite. In her zeal to get things right, Cantrell had written to the Smithsonian for information about the fragment of the Pitts meteorite in its collections. Cantrell explained that she had spoken with a Smithsonian curator on the phone, who was "one of the nicest people she had ever met." Linda Schramm, who was the collection manager of the Division of Meteorites at the time, provided the photographs. She has since retired, but I was proud to hear of her willingness to help a woman she did not know who lived in Pitts, Georgia.

I recorded Ruth Cantrell's story about her mother's experience with the meteorite, during which she said:

> My mom did see the meteorite when it fell. She was out in the yard at her parents', the Melville Wood family, when she was scared to death by what she thought was a ball of fire coming out of the sky. She thought the Lord was coming, but she said she wasn't ready, and she started screaming and carrying

> on. It so happened that my grandmother Lewis's mother was visiting with them at the time, and she heard my momma screaming. And she came out on the porch looking up at the sky and fell off of the porch.

About her mother's recollections, Cantrell added, "She did not see it land, but she saw it come and go over her house, and it was evidently making a noise, and that was why she saw it. It landed over in a field near Mac Pearce's house."

Cantrell indicated that her mother's home was two or three miles to the west of her present home, which would have put it near the center of Pitts. That her mother saw the meteorite pass over her house and knew that it landed to the north of town was consistent with the observation made by McCallie that the meteorite was traveling from south to north.

Before we left the Cantrells, Dock asked Ruth to play a tune he liked on the piano. She demurred at first, saying she probably could not remember it, but she finally sat down, opened the piano keyboard, and began to play. Initially, she was a bit hesitant, and the notes came slowly, but then it all seemed to come back. I was taken by surprise because my Southern Baptist upbringing led me to expect something a bit mournful, but this tune had a catchy syncopated rhythm, something between ragtime and old time religion. I still have my recording of her performance and smile when I listen to it. I can see why Dock likes it.

As we thanked them for giving us their time, I knew we had met two American originals—people who had made the most of their lives.

### *Final Stop: Osco Bennett*

Our last visit was to a nursing home in nearby Ashburn in Turner County where Osco Bennett, 102 years old, was a resident. As a boy he had witnessed the passing of the Pitts meteorite over his home. He was waiting for us in his room with his son and daughter-in-law.

Mr. Bennett was in good spirits. At his age he had plenty of memories to draw on, and he enjoyed the opportunity to share them. The challenge was to get him to realize that I was only interested in a small part of his long life. To him, the meteorite story was something that happened one day and was pretty much over after that, but to me it was so much more. Fortunately, his son was there to remind him or encourage him with questions to tell us about that day.

Although I recorded the hour-long interview, it is difficult to quote Bennett at length because his recollections of the meteorite were woven in and among other memories of his life. The gist of his story is that he

Osco Bennett, 102 years old, is interviewed by the author at a nursing home in Ashburn, Georgia.

was "born and raised in Wilcox County," and as a boy he lived in a house located on the north side of Pitts. He said his family's home was about "halfway between town and the Snow Hill Church," which was the African American church we had visited earlier in the day, the likely landing site of the fragment of the meteorite now in the Smithsonian collections. Bennett recalled that four or five meteorite fragments were found and said that the one that went over his home fell in the yard of the Snow Hill Church, which was consistent with what we had learned from other sources.

He was a bit unsure about the date when the meteorite arrived, mentioning both April 20 and April 21 as possibilities, but he leaned toward the twentieth, which is the right day. He said the morning was clear and sunny with no sign of a cloud in the sky. He was in the yard, playing a game of marbles with a friend—a thought that elicited other memories on the topic of marbles. It was clear from his stories that Bennett had enjoyed playing marbles. He went on to say that while he was playing with his "black, shiny marbles" about 9:30 a.m., "boom, boom, boom, a series of explosions went off directly overhead."

He said he was not afraid but was excited to hear something like fireworks, and he looked up to see the cause. When asked, he said he did not see the fireball itself but "I looked up in the air from where the sound came and there was a big black bubble, and a little higher over where the explosion had already happened, the black smoke was piling up, rolling, spreading on out." He continued:

> In a matter of minutes, I would say three or four minutes, here come another big blast and another one succeeding that. . . . Big black smoke boiling up, and rolling, you know, getting bigger and bigger and bigger. A whole series of explosions had passed, oh, I guess two or three minutes maybe; it wasn't just a boom, boom, boom, boom, but boom one time, and wait, silent, and boom again. It was all in succession. And the [smoke] blossoms, I like to think about it like when you see flowers opening up, you know. And I consider that this was the first bomb that went over. Then it brightened up to a full rolling sound, and then about a few minutes later here come another sound and the blossoms were going in all directions.

When I first heard this narrative, it seemed that he might have been confused, but I later discovered that his story was consistent with observations of other meteorite falls. In the "Fireball FAQs" by the American Meteor Society, three of the questions are pertinent to Bennett's story: "Can a fireball create a sound?," "Will the sound occur right away as you watch the fireball, or is there some delay?," and "Will a fireball leave a trail?" The answers depend on the conditions at the time of observation, including time of day, weather, and angle of meteorite path. For conditions like those when the meteorite passed over Pitts, the FAQs say, "There is [a] chance of sonic booms. . . . it will generally be 1.5 to 4 minutes after the visual explosion before a sonic boom can be heard [since sound travels much slower than the meteorite]. Observers who witness such spectacular events are encouraged to listen for a full 5 minutes after the fireball for sonic booms." I had to chuckle over this advice, since there is little chance that anyone will be a witness to a meteorite fall of this nature, and it is even more unlikely that they will have prepared for the event by reading this helpful tip beforehand. Yet Bennett, playing marbles at the time, was perfectly positioned to be the ideal witness of a meteorite fall.

Bennett's recollections about the smoke created by the meteorite are also addressed in the FAQs: "Fireballs can develop two types of trails behind them: trains and smoke trails. The second type of trail is . . . more often seen in daylight fireballs than at night. Generally occurring below 80 km [50 miles] of altitude, smoke trails are a non-luminous trail of particulate stripped away during the ablation process. These appear similar to contrails left behind by aircraft and can have a dark appearance." This information fits Bennett's circumstances and his description of what he saw.

What really caught my attention was that what Bennett witnessed required conditions that had to be just right—in bright sunlight and observed at a point in the fall where the meteorite is moving at an acute angle before it hits the face of the earth. Both of these criteria were present for his observation of the Pitts meteorite. His was a rare experience.

With my interview of Bennett approaching an hour, we could see that while he was still excited, he was getting tired. We called it quits and bid farewell to Bennett and his helpful son and daughter-in-law. Thus ended a memorable day. I had met a group of remarkable people and learned more than I thought possible—and all because of a mistake in the scientific literature.

## Tying Up the Loose Ends

South Georgia's geology may not have the allure of places with mountains or earthquake faults, and no rich lodes of gold or silver are found there, but the place is interesting in its own way. The Smithsonian collections, with their breadth and variety, provided me with specimens that connected to stories I never expected before I made my visits to Enigma and Pitts. Of course, it was a cherry on top when Frankie Snow was able to fill a gap in the collections by providing three specimens of Altamaha grit.

As always, there were interesting and helpful people attached to the specimens in the collections, particularly the meteorites. The Smithsonian curator Cari Corrigan was a fount of knowledge and a patient teacher for an old student trying to learn something new. Because of a mistake in the scientific literature, I met some remarkable folks: Mike Miller, the man who knew about the mistake, and Mac Pearce, Ruth Cantrell, and Osco Bennett—three people whose memories helped to correct the mistake. Each was delightful.

I sent my findings about the Pitts meteorite to Corrigan, who said she would make sure that the meteorite records reflected what we had learned. As for me, I was content with a search that had begun in the Smithsonian collections and led to Pitts, Georgia, and finally to the right answer about the landing of a meteorite.

CHAPTER TEN

# Early Native American Civilizations

## *A Long Story Cut Short*

The history of the place we now call South Georgia is a story that goes back millions of years. Within this long time frame, humans arrived only a short while ago, sometime in the past fifty thousand years. Most of what we know from recorded history concerns just the last five hundred years or so. To learn about the long story prior to recorded history, we turn to archaeology. And when it comes to archaeology, the Smithsonian has collections second to none. My interests were in what these collections had to say about the peoples who hundreds and thousands of years ago lived in the area that would become South Georgia. This task opened a window on a world that was far more complex than I had imagined.

My search was complicated by the fact that the Smithsonian has not just one set of archaeological collections, but two. The first set is housed in the Department of Anthropology of the National Museum of Natural History. These are among the oldest collections at the Smithsonian, with some artifacts dating back to the Wilkes expedition, which circumnavigated the globe before the Smithsonian was founded. Representing archaeology, ethnology, and physical anthropology, these collections hold more than 3.1 million objects and 14,000 cubic feet of archives. The archives by themselves include more than a million photographs, 10,000 sound recordings, 8 million feet of film and video, and 20,000 artworks. Most, but not all, of the artifacts were documented on acquisition following scientific protocols.

A second set of collections is housed in the National Museum of the American Indian, founded in 2004. More than 70 percent of the collections of this museum came from one man, George Gustav Heye. Born in 1874 to a wealthy family in New York, Heye's first job after earning his engineering degree in the late 1800s was in the West, where he began collecting artifacts from the indigenous people. Over a period of fifteen years his collecting hobby became his life's work. The scope of his interest expanded to Native American cultures ranging from Alaska and Canada to Chile, and in 1915 he even worked in northeastern Georgia with a team excavating the Nacoochee Mound. Eventually he created his own museum in New York City to display his collections, and he never stopped collecting until his death in 1957. His collections grew so fast and became so extensive that the artifacts were often poorly identified as to source or character.

After Heye's death his museum fell on hard times, and in the 1980s a group of thoughtful people, including David Rockefeller and Senators Patrick Moynihan of New York and Daniel Inouye of Hawaii, came to the Smithsonian with a proposal to create the National Museum of the American Indian. The new museum would take over the Heye collections and begin the long process of establishing documentation on the background of its artifacts. Since the founding of the museum, its collections have grown to encompass 825,000 objects, including 324,000 photographic images. The work to document the Heye collections still continues as I write this book.

My endeavor to explore the collections for South Georgia artifacts was aided by archaeologists from the National Museum of Natural History, including Eric Hollinger, Torrey Rick, Dorothy Lippert, and Bill Billeck, and from the National Museum of the American Indian, Jane Sledge and Ann McMullen. Both sets of collections hold artifacts of interest to my task, but the collections in the Department of Anthropology in the Museum of Natural History proved to be more useful because they hold more objects related to South Georgia and have more supporting documentation.

## Sorting Out the Variables

At the outset of my search I knew that in the distant past Native Americans had built large mounds in Georgia. The most prominent are at the Etowah Indian Mounds Historic Site in North Georgia and at the Ocmulgee National Monument and the Kolomoki Mounds State Park in South Georgia. The mounds at what we now call the Ocmulgee National Monument inspired William Bartram to write in *Travels through North and South Carolina, Georgia, East and West Florida*, "On the East bank of the river [Ocmulgee] lie the famous Oakmulgee fields, where are yet conspicuous and very wonderful remains of the power and grandeur of the ancients of this part of America." These mounds are impressive, but they are only one manifestation of early Native American cultures and were built within the last thousand years. The history of human occupation in the southeastern United States extends much further back than this time, at least thirteen thousand years earlier and perhaps much more.

I was confident that the collections would contain artifacts from the great mounds in Georgia because the Smithsonian conducted excavations there. But would the artifacts speak of the people who lived in South Georgia before the mound builders? Would the artifacts tell the story of the inhabitants during the days of the megafauna ten thousand to thirteen thousand years ago? Was there evidence for societies that might have existed even before that? This search enmeshed me in mysteries, debates, and discoveries.

## Who Were They?

The questions of when the first humans arrived in North America, how they got there, and how they spread across North and South America are live wires in archaeological circles. Successive theories about how this journey occurred have each eventually fallen victim to evidence from new discoveries. At one point in the late 1800s and early 1900s, it was believed that human occupation of North America could not have occurred earlier than ten thousand years ago. Today, according to Tim Flannery, the most widely accepted narrative for the arrival of humans in what is now the southeastern part of the United States begins with the passage of people from Asia to Alaska fifteen thousand to seventeen thousand years ago, when low sea levels associated with the last glacial age exposed a land bridge across the Bering Strait. The narrative then has the descendants of these migrants making their way down the western coast of Canada. When they had gone far enough south to skirt the still-existing continental glacial front, they turned east. The current assumption is that this migration continued east until the first humans arrived at the Atlantic Ocean about thirteen thousand years ago.

However, studies in the eastern United States continue to push back the date of human occupation. Work at the Ocmulgee Mounds in South Georgia suggests that humans occupied the site perhaps continuously for more than seventeen thousand years. Archaeologists from the University of South Carolina studying the Topper site on the Savannah River have evidence of human occupation going back twenty thousand years or perhaps even more. These and other findings raise questions about the Bering Strait crossing theory. One of the challenges to understanding more about Georgia's earliest residents is that twenty thousand years ago the earth was still in the grips of an ice age and the sea was four hundred feet below present levels. If early people lived near the coast, the sites that might tell us more are now under several hundred feet of water.

Still, archaeologists continue to look for evidence, and one clue associated with early humans in North America is the Clovis point. In 1932 a road crew working near Clovis, New Mexico, unearthed a cache of fossil-like bones. Edgar B. Howard, an archaeologist working at the University of Pennsylvania Museum of Archaeology and Anthropology who had a long interest in such finds, was notified. Within days he arrived and made preparations for an archaeological investigation. What he found was not just the bones of a mammoth, but within the skeleton there were a number of long spear points that had been used to kill it. These points would become known as Clovis points. The mammoth remains were dated as thirteen thousand years old, which was three thousand years before the then-accepted date for the arrival of the first humans in North America. Howard was clever enough to leave the scene untouched to allow other archaeologists to visit the site and verify what he had found. While the evidence was conclusive, it would take a few more years before it was accepted that people had not only lived on the North American continent thirteen thousand years before, but also had used spears to kill large animals.

In the decades following, Clovis points were found in the Midwest and the Southeast. As these findings accumulated, another orthodoxy of human occupancy was born: the Clovis first theory. The Clovis first theory assumed that the first humans had crossed the Bering Strait earlier than previously thought. The theory postulated that once established, people began to migrate eastward across the continent, carrying Clovis points with them. But questions about the Clovis first theory began to arise as sites were discovered in the eastern United States that suggested human occupation earlier than thirteen thousand years ago. As the evidence of early sites increased, agreement emerged that pre-Clovis humans must have existed. Given all of the variables still in play, it remains unclear who the pre-Clovis people were and how they got here.

I have held Clovis points in my hands that date to thirteen thousand years ago—beautifully flaked out of hard chert or obsidian. Attached to a spear, these points could be used to kill a large animal. Over time, as the larger prey animals were eliminated, Clovis points became smaller. Eventually they evolved into what archaeologists refer to as Dalton points. The Smithsonian collections contain examples of both of these projectile points found in South Georgia.

As interest in archaeology in the Southeast has grown, more Clovis and Dalton points have been found in Georgia. With their discovery, the Smithsonian reexamined its pre-1930 collections from the region and found they contained Clovis and Dalton points. The growing numbers of Clovis and Dalton points being found in South Georgia suggest that much is yet to be sorted out. New archaeological finds and fresh insights into old collections continue to help fit together the stories of ancient cultures like a puzzle, piece by piece, steadily shaping and reshaping our understanding of who lived on the land in generations past and what debts we owe them.

To understand the time frames associated with the peoples who made these weapons and other objects of antiquity, archaeologists have constructed timelines associated with different cultures. Some of them are complicated, but here is a simple one:

Pre-Paleoindian period: 18,000 BCE–12,000 BCE
Paleoindian period: 12,000 BCE–8,000 BCE
Archaic period: 8,000 BCE–1,000 BCE
Woodland period: 1,000 BCE–900 CE
Mississippian period: 900 CE–1540 CE

Within each of these periods, archaeologists have subdivisions (like the Early, Middle, and Late Woodland periods), but for the most part that level of detail goes beyond the purposes of this book.

Using this breakdown, the people who used Clovis points lived in the Pre-Paleoindian and Paleoindian periods. The people who built the great mounds were of the Mississippian period. The current theory for the arrival of mound builders in Georgia is that they came from the West around 900 CE, bringing with them large-scale cultivated agriculture. While the Mississippians became dominant after their arrival, elements of the Woodland culture that preceded them survived and were in some cases adopted by the Mississippians. However, when Christopher Columbus arrived in the New World in 1492, it did not take long for everything to change.

## Where Did They Go?

This question as it relates to South Georgia takes us to the National Portrait Gallery, whose collections contain an image of a Spanish conquistador named Hernando de Soto. De Soto's portrait is actually an engraving, made in 1865 from an original portrait. In the engraving, de Soto is in armor with a sword at his side, looking very much like the soldier he was. His hair is short and curly, his nose aquiline, but it is his piercing black eyes that strike the viewer. People feared de Soto not only because of his withering gaze but also because of his history. He had come to the New World around 1520. He gained notoriety for his role as a captain with Francisco Pizarro in the brutal conquest of the Incas in Peru, learning early about killing and enslaving native people.

De Soto experienced success in Peru with Pizarro and brought untold riches back to Spain, but he was looking for an expedition of his own making. This desire led him to North America in 1539, when he landed his fleet near what is today's Tampa, Florida. He moved his force of more than six hundred men and women to a site near Tallahassee, where they overwintered. In 1540 they began a journey that passed through Georgia, continued northward into North Carolina, turned west all the way to Texas, and eventually ended along the banks of the Mississippi River. We know from a journal of an expedition member (*The De Soto Chronicles*, edited by Clayton, Knight, and Moore) that he passed through South Georgia following the rivers of the area, but his exact route has not been established. In the 1930s a Smithsonian archaeologist found evidence of a rocky shallows where de Soto was thought to have crossed the Oconee River. More recently, evidence of his presence has been found at a site on the Ocmulgee River, not too far from my grandfather Clough's old farm. Originally identified by Frankie Snow, this site has been excavated by a team from the Fernbank Museum of Natural History in Atlanta. The location has been documented as an established indigenous village, but also found there were metallic objects of Spanish origin that fit the profile of those used for trade by de Soto.

What we know about de Soto's expedition is that he brought with him not only six hundred people, but also two hundred horses and a host of hogs. He found none of the gold and silver he was seeking and was harassed along the way by Native Americans who had learned he was not to be trusted. After meandering across the Southeast for most of three years, he died in 1542 after reaching the Mississippi River with little to show for his efforts. However, the diseases his expedition had introduced, including smallpox, had devastating and lasting effects.

Some estimates say that up to 95 percent of the indigenous people of the New World died from diseases contracted from early contact with Europeans, and the de Soto expedition was one of the sources. The actual number of people who died in this manner is a matter of dispute, because large indigenous population centers were subject to declines from causes unrelated to

Europeans. A view I find more plausible is that even if the European-induced pandemics did not kill as many people as previously thought, the effects were still drastic enough to stun the indigenous communities, playing a major role in the collapse of their cultures. (Losing 50 or 60 percent of the population of a culture today would destroy it.) We also know that the next Spanish expedition to reach the area in 1560 reported much smaller populations of Native Americans than were seen by de Soto.

Yet another theory is that de Soto's hogs carried bacteria harmful to humans, worsening the impact of the expedition as native people killed and ate those that had gone feral.

We still do not fully understand the rapid decline and disappearance of the early Native American civilizations in South Georgia, but we do know it coincided with the arrival of Europeans. Remnants of the indigenous peoples of South Georgia eventually came together to form what would be known as the Creek Confederacy or Muscogee Nation.

## The Division of Mound Exploration in Georgia

Any discussion of Native American artifacts at the Smithsonian has to begin with the Division of Mound Exploration, which was formed in 1881 at the request of Congress. During the 1870s, a heated national debate had arisen about who had built the hundreds of ancient mounds found throughout the Midwest, Southeast, and Middle Atlantic. Stories abounded about builders who could have come from Europe, the Middle East, or Central America. The debate was so intense that Congress asked the Smithsonian to settle the matter. Cyrus Thomas, a seasoned ethnologist, was chosen to head the new division, and he set about building a staff of archaeologists to carry out the work. When it was all over in 1894, more than two thousand mounds had been explored, and more than forty thousand artifacts had been added to the collections of the Smithsonian.

Thomas faced many issues, including determining which of the thousands of known mounds to investigate and how to deal with the hundreds of questions that passionate amateurs thought needed resolution. Recognizing the scope of the task he faced and his limited resources, Thomas was choosy about which mounds he would study. He was also careful to focus the mission of the work by stating, "The most important question to be settled is, 'Were the mounds built by Indians?'"

To apply his resources to best effect, Thomas decided that if a credible investigation had already been conducted by someone else, he would accept that work and focus his efforts on other sites. When it came to Georgia, he knew that a brilliant and somewhat eccentric historian, Charles C. Jones Jr., had already conducted a groundbreaking study of early Native American cultures, including South Georgia mounds, such as those at Kolomoki. Jones was classically educated at Princeton and Harvard and served as mayor of Savannah before the Civil War. He was a plantation owner and fought in the Confederate army; he was financially ruined after the war. He moved to New York to reestablish a law practice, which thrived, but he eventually returned to the South and became known as the foremost Georgia historian of the nineteenth century.

Most important to Thomas was Jones's book *Antiquities of the Southern Indians, Particularly of the Georgia Tribes*, in which Jones documented previous investigations of Georgia mounds, as well as his own studies. The book not only was recognized for its rigor but also established the protocol for future archaeological studies. Eventually a few of the specimens from Jones's work ended up in the collections of the Smithsonian.

## The Etowah Mounds

Because of the work of Jones, Thomas only investigated a limited number of sites in Georgia, with the most extensive work being done at the Etowah Indian Mounds Historic Site near Cartersville. Etowah lies to the north of Atlanta, so it does not fit my South Georgia profile, but the people who built those mounds had trade and cultural connections with people living in villages down the Etowah and Savannah rivers. A brief introduction to the Etowah Mounds is useful to understanding these downstream cultures, which were the source of artifacts obtained in South Georgia.

Thanks to more recent archaeological studies, we know that the Etowah Mounds were not built in one fell swoop and were not occupied continuously. While the site writ large had been occupied well before the mounds were built, the mound builders created an important village there and occupied it off and on from around 950 CE to 1450 CE. The mounds we see today were built over time as the site was occupied by successive generations. The location was conducive to maintaining a large population, with fertile river plains where crops could be grown and proximity to the Etowah River and to the mountains of Georgia, which offered abundant animals to hunt. The number of people at the Etowah site ebbed and flowed depending on the food supply, the weather, enemy attacks, and disease, but even after a disaster struck, people would later return and reuse the site. Some mounds served as ceremonial edifices and others as burial places for leaders and their families.

The Smithsonian Division of Mound Exploration began work at the Etowah Mounds in 1884, and it yielded impressive artifacts, including two unusual sculpted copper plates depicting a mythical birdman figure holding a mace in one hand and a severed head in the other. The plates were named after the archaeologist John Rogan, who discovered them. Before I retired from the Smithsonian, it was my good fortune to view one of the Rogan plates. Although it suffered from its time in the ground, the birdman seems ready to dance off the copper surface. The image reflects the worldview of the Mississippian culture. Also known as Morningstar to historic tribes in the Southeast, the birdman is a heroic figure who defends living people from beings from the Beneath World. In the tale, Morningstar loses his life in a battle with the Beneath World, but he is avenged by his sons and finds new life as the morning star in the sky. This beautiful parable has parallels in stories from other religions.

The Rogan plates are also of interest because they are made of copper, which is not found in the Etowah area. Archaeologists have proposed that the copper was probably mined near the Great Lakes and shaped into the plates at the Mississippian center of Cahokia in what is now Illinois. This theory suggests that the Etowah people participated in a trade network that was hundreds of miles long, and it may indicate that some of their number had migrated to Etowah from Cahokia. This same culture also migrated down the rivers from Etowah and into South Georgia.

## The Hollywood Site

As the work of the Mound Division in Georgia was winding down, Cyrus Thomas learned of one more opportunity—a lesser-known site near Augusta named the Hollywood Mound site—and he encouraged one of his best young archaeologists, Henry Lee Reynolds Jr., to study it. While the Hollywood Mound site was not rich in numbers of artifacts, those that were found provided rich insights. Reynolds's study was also important for my work, because the Hollywood site is in Richmond County, south of the fall line and hence in South Georgia.

Henry Reynolds, known as "Harry" to his friends, was born in Washington, D.C., in 1861 and earned a bachelor of arts and a law degree from Columbian College (later renamed George Washington University). His life was chronicled by Mark Williams and Keith Stephenson in an article titled "Henry L.

Reynolds, Jr.: Georgia's Greatest 19th Century Archaeologist." High praise for a young man who died at the age of twenty-nine, just two years after his work at the Hollywood site was completed.

Reynolds's education seemed to predestine him to practice law, but by 1885 he was in the employ of the Smithsonian Division of Mound Exploration. He took on investigations of increasing complexity, and in 1889 Cyrus Thomas assigned him to lead the exploration of the Hollywood Mound. The Reynolds report was so exemplary that Thomas included it almost verbatim in the summary on the work of the Mound Division.

Reynolds reported, "There are two mounds situated in a bend of the Savannah River in Richmond County, Georgia, three miles east from Hollywood, a small flag station on the Georgia Central Railroad about ten miles below Augusta, and five miles above Silver Bluff." Both mounds were modest in size and located in a floodplain of the Savannah River. The site was later farmed, and the surfaces of the mounds were modified from their original state. Reynolds only excavated Mound 2, the smaller of them, because he believed it was a burial mound, and he was right. It contained objects that provided insights into the culture of the people who had built the mounds and their relationship to others living in Georgia, South Carolina, and elsewhere in the Southeast.

Of the two levels he found in Mound 2, he wrote, "In the lower division . . . , all the articles seemed to be clustered around the hearth. . . . Here most of the human remains were found, but, like those in the upper burial, only the merest traces were observed. . . . Decayed and meager as they were, sufficient evidence was had in the case of each skeleton to show that it was human."

Reynolds sent to the Smithsonian ceremonial pipes, carefully made stone disks used for the game of chunkey, pottery, fragments of embossed copper plates, large inscribed pots, and human remains. On one of my visits to the Smithsonian archaeological collections, I made it a point to see the Reynolds artifacts. Wearing gloves, I examined the beautifully patterned pottery, the elegant pipes, and the remarkable embossed plates. As I marveled at the artisanship of the objects, I could not help feeling that I was invading the privacy of the people, who never expected that their personal objects would be taken from their graves and end up in my hands.

One of the copper plates from Mound 2 is intact enough to exhibit a clear embossed image of what is known as the underwater panther. Like the birdman on the Etowah Rogan plates, the underwater panther is a mythological creature associated with the Mississippian and other indigenous

Earthen bottle, Mississippian, Hollywood site, collected by Henry Reynolds, 1889. National Museum of Natural History anthropology collections.

Copper plate, underwater panther, Mississippian, Hollywood site, collected by Henry Reynolds, 1889. National Museum of Natural History anthropology collections.

cultures. The underwater panther inhabits the world under land or under water, which in this Native American belief system is a place of chaos but also offers the possibility of redemption and renewal.

In the years since Reynolds conducted his excavations, several new studies have been carried out at the Hollywood site, the most recent under the auspices of the Savannah River Archaeological Research Program, the South Carolina Institute of Archaeology and Anthropology, and the University of South Carolina. Based on the results of ground-penetrating radar and on Reynolds's maps, the precise original locations of his artifacts were identified.

Thanks to Frankie Snow and his colleagues the archaeologists Chris Thornock and Keith Stephenson, I was able to visit the site while it was under investigation. Because Reynolds did not have access to today's knowledge, he did not fully appreciate the implications of his finds. Research has shown that while the burial urns are of local origin, the other objects were imported from as far away as Tennessee and Mississippi. Most of the artifacts can be linked to similar finds from the Etowah Mounds. These objects and the dates established for the site have led researchers to conclude that the mounds at Hollywood were built when the Mississippian culture was taking hold over the local culture around 1300 CE. The Mississippian influence would prove ephemeral, however, since by the early 1400s the Mississippian culture was in retreat, consolidating to the west and north.

Thanks to the work of Henry Reynolds and those who have followed, we know that the culture at Hollywood was a blend of old and new, and it was connected to the larger world through trade and exchange. It is clear that the people who lived there valued aesthetics and had their own belief system. We only know them through a few artifacts, which tell us they appreciated self-adornment, enjoyed games, and were fearful about what they did not know but hopeful for the future. A lot like my family.

In 1894 the work of the Division of Mound Exploration was formally concluded by the submission of a massive report by Cyrus Thomas to the Smithsonian Board of Regents; this report is now considered a landmark in American archaeology. While there were hundreds of discoveries, there was one major finding: the mounds were built by the people living in the vicinity a couple of hundred years before Europeans arrived.

## The Ocmulgee National Monument Mounds

Although Cyrus Thomas and the Division of Mound Exploration did not explore the mounds at South Georgia's Ocmulgee National Monument, the Smithsonian got a second chance, thanks to the Great Depression. Three forces worked together to create the opportunity: local citizens, who were concerned the site would soon be lost to development; the new administration of President Franklin D. Roosevelt, which was willing to fund the effort upon learning it would create hundreds of jobs; and the Smithsonian, which was willing to undertake the archaeological work.

The Smithsonian was authorized by the federal government to excavate the Ocmulgee Mounds in 1933 as part of a public organization known as the Civil Works Administration. When the better-known Works Progress Administration (WPA) was created in 1935, it took over the project. The goal was to employ not only archaeologists but also hundreds of men and women who could help with the physical labor of the excavation and sorting the artifacts. Arthur R. Kelly, a young classically educated Harvard anthropologist, was selected to lead the project for the Smithsonian.

In 1936, with the work well underway, Roosevelt signed legislation designating the site as the Ocmulgee National Monument, and the National Park Service became the administrative entity for the site. Kelly was appointed superintendent of the Ocmulgee National Monument by the National Park Service, and the role of the Smithsonian was gradually reduced. As a result, the collections from the work, which were enormous, were split between the Smithsonian and the park service, with the largest share going to the park service.

With upward of eight hundred WPA employees digging at the site, the collection of artifacts grew rapidly. In the end, the flow of objects overwhelmed the archaeological staff, so that even today many pieces are not identified. In spite of such shortcomings, the work done during the WPA period was invaluable to our understanding of the different indigenous cultures that had occupied the site. Artifacts from the Mississippian mounds include elegant pottery, large and small ceremonial smoking pipes, and projectile points.

One of the signature finds outside of the main mound complex was made by Kelly in 1938. It was a form of pottery now known as Swift Creek, named for the stream adjacent to the location where it was first discovered. Swift Creek pottery documents the presence of Woodland people at the site and thus opened a new thrust for research.

Kelly remained at the site until 1939, when he was promoted to chief archaeologist of the National Park Service. By 1942 the WPA program had ended, and the large-scale investigation at Ocmulgee did as well. In the end the work uncovered more than two million objects.

## Private Collections

The work of the Division of Mound Exploration in the late 1800s created an awareness that the Smithsonian was in the "Indian artifact business." Soon, private citizens were offering to sell or donate their Native American collections to the Smithsonian. To find any that came from South Georgia, I began with the digital collection records of the Smithsonian. I entered "Georgia" and "arrowheads" as keywords, and I was pleased when 110 listings appeared. Each showed the county and/or site name, collection date, and the collector(s) or donor(s) if this information was available.

The first listing indicated that the artifact was a bundle of eleven arrowheads and one knife point found in Columbia County near a site called Steiner Place. The collector was Roland Steiner. I felt I was getting somewhere, since Columbia County is on the Savannah River south of Augusta and falls within my boundaries for South Georgia. A digitized handwritten document showed the donation had been made in 1896.

The collector for the second listing was also Roland Steiner, and he showed up again on the third one. I had to go all the way to page nine of the listings to find a collector other than Roland Steiner. The collector listed in that case was Jonathan McGlashan, but no county was given for his contribution. McGlashan's name, unlike Steiner's, did not repeat anywhere in the remaining listings. I later discovered that the lack of his presence in the records was a bit misleading. Steiner, however, was clearly significant.

A bit of research revealed that Roland Steiner was a Civil War veteran, a medical doctor, a wealthy planter, and an amateur archaeologist. He was a founding member of the American Anthropological Association and had a long and productive relationship with the Smithsonian. The Smithsonian Archives contain more than 250 letters and memos written either by Steiner or by Smithsonian curators to him between 1891 and 1904. They cover transactions in which Steiner provided nearly eighty thousand Native American objects to the Smithsonian collections.

I was fortunate to learn of the work of Daniel T. Elliott, a scholar who is not related to the Smithsonian but who studied the Steiner collections at length. Elliott had not only reviewed the same records I saw but also deciphered Steiner's handwritten letters, which was a challenge to say the least. Elliott's work and the archival documentation show that the acquisitions from Steiner were purchases made when Smithsonian budgets allowed. Apparently, the initial intent was to fill in the gaps from the Smithsonian's own Mound Division excavations at Etowah by acquiring objects Steiner

Discoidal (chunkey) stone, Mississippian, collected by Roland Steiner in Richmond County, Georgia, September 1900. National Museum of Natural History anthropology collections.

had obtained from Mound C. But with time the Smithsonian connection with Steiner grew to include large numbers of objects he obtained from excavations at sites in Columbia and Burke counties near his South Georgia home. As the years passed, Steiner wrote letters to the Smithsonian brimming with enthusiasm about his discoveries.

Steiner followed professional protocols in his work and provided documentation for his finds as best he could. However, he was not able to clarify the history of his objects in all cases, and at the time of his discoveries, little was known about what types of objects fell into which time periods. As the artifacts poured into the Smithsonian, a backlog grew in the efforts to provide finished documentation.

According to Elliott, the Smithsonian's Steiner collection eventually fell on hard times. In 1906 Steiner died, and the archaeologists who had worked with him eventually departed, leaving behind the daunting task of sorting his vast collection. Matters came to a head in the early 1930s during the Great Depression. Facing budget constraints, the Smithsonian decided it could no longer keep large collections that were not being actively used or that lacked sound documentation. A significant portion of the Steiner collection fell into this category, and many artifacts were disposed of or loaned to other institutions or to schools as educational materials.

I would be remiss if I concluded this discussion of private collections without returning to Jonathan McGlashan, whose name I encountered during my initial digital search of the anthropology collections. At the time McGlashan seemed to be a minor player, but that did not prove to be the case. I am indebted again to Daniel Elliott and to a young colleague of his, Elizabeth "KC" Jones, for background on McGlashan.

As opposed to Roland Steiner, McGlashan was a man of the working class, born into a family of railroaders in Scotland. He migrated to the United States and was hired in the mid-1840s by the Central of Georgia Railway, which provided rail service between Savannah and Macon and points west. During his early days as an employee, he was engaged on a project to relocate tracks to a less flood-prone route through the Ogeechee River valley. There, McGlashan found his first indigenous objects and began a collection. Although rail service in this area was disrupted by Union troops on Sherman's march to the sea, it was restored and operable again in 1866. McGlashan continued to work with the Central of Georgia Railway until his death in 1884. Over his lifetime he amassed a collection of about twenty thousand Native American objects, most of which he unfortunately did not document as to source or culture.

In 1886 McGlashan's widow, Mary, offered the collection to the Smithsonian for purchase. In a memorandum written by Cyrus Thomas to then Smithsonian secretary Spencer Baird, he stated, "I send you the McGlashan collection, which as you probably remember is from a lady in Georgia. The collection contains by actual count over 19,000 pieces, and will be quite useful to you for purposes of exchange. As the lot cost only $250.00, I think it is very cheap." Thomas's memo indicates that at the outset the collection did not have great scholarly value, and he foreshadows its dismemberment through "exchange."

While McGlashan's collection potentially offered a resource for the study of Native American cultures, the lack of documentation and the large size of the collection were a problem. As a result, the McGlashan collection was gradually dismembered to an even greater extent than the Steiner collection was.

If there is a silver lining to the story of Steiner and McGlashan, it is that the Smithsonian no longer acquires collections unless they can be cared for and held for future use. What remains of the Steiner and McGlashan collections has now been stabilized and protected. A new appreciation is developing for these artifacts as a source of information about the early human occupations of South Georgia. With an increasing number of Clovis and Dalton points being found there, for example, the Steiner and McGlashan collections have been reexamined and found to contain these weapons. In addition, as more of the objects are digitized and made available to the public, more scholars and enthusiasts are likely to access them.

## Swift Creek Pottery

The discovery of Swift Creek pottery by A. R. Kelly in 1938 at the Ocmulgee National Monument introduced the world to a culture that flourished there before the Mississippians. It is believed that the people who made the pottery lived in the Woodland period, probably from about 100 BCE to about 900 CE. Their culture centered in South Georgia but extended above the fall line and into Alabama and Florida, but their lives were upended as the Mississippians moved east. The Mississippians built large mounds and depended on cultivated crops, while the Swift Creek people were largely a hunter-gatherer society with only limited cultivated agriculture. Their villages were moved more or less annually to improve hunting and access to other food sources.

Swift Creek pottery from South Georgia, largely in the form of sherds, is found at the Smithsonian in the collections of the National Museum of the American Indian and the National Museum of Natural History. The Ocmulgee National Monument also has collections, and my friend Frankie Snow has assembled his own formidable collection from more than four hundred sites he has discovered in South Georgia.

Swift Creek pottery is known for the patterns created by skilled potters using paddles to stamp images onto the pottery before it hardened. Technically it is known as Swift Creek Complicated Stamped pottery. Other early Native American cultures also made pottery with images or patterns, but the Swift Creek people were masters at creating intricate patterns that are pleasing to the eye. The images often reflect animals indigenous to the area or are abstract with curvilinear patterns. Over the hundreds

Frankie Snow and the author with Snow's collection of Swift Creek pottery and paddle images, South Georgia State College.

of years that such pottery was made, the Swift Creek patterns changed, and other cultures adopted variations of them. Yet Frankie Snow and others have shown that the Swift Creek patterns were unique and served as a marker for their culture. Frankie has identified some 450 distinct patterns in the paddles the Swift Creek potters used.

Published in 1998, *A World Engraved: Archaeology of the Swift Creek Culture*, edited by Williams and Elliott, provides a summary of the findings made between Kelly's discovery of the pottery and the year of publication. In a paper in this book, Frankie advanced the idea that the Swift Creek patterns were more than decorative; they were windows into the people and the culture: "Potentially, this prehistoric artwork could tell us much about the people who made Swift Creek pottery: it could yield information about these people's world view, magic, religion and folktales not accessible by other archaeological analysis." His use of the word "potentially" was an understatement, because by 1998 he had already documented dozens of examples that illustrated his point. When I visited his collection in 2014, Frankie had taken his work well beyond his 1998 paper to show the connections that Swift Creek potters made with natural and supernatural beings. He was honored in 2014 with the prestigious Crabtree Award from the Society for American Archaeology for his work on Swift Creek pottery. This award is given to an individual who is not formally trained in archaeology but has made significant contributions to the field.

For all of the progress made in understanding the Swift Creek culture, much more is yet to be discovered. After all, less than a hundred years have passed since A. R. Kelly first identified the pottery. What we do know is that the people of the Swift Creek culture were hunter-gatherers who had a deep understanding of the connections between the natural world and humans.

## Striving to Understand the Sweep of History

Swift Creek pottery offers insights into a culture that dates back a thousand years, but what of the Archaic or Paleoindian periods, which go back thirteen thousand years or more? Here the evidence is skimpy but enticing. Little remains to document the presence of these cultures other than Clovis and Dalton points.

The artifacts obtained by Smithsonian archaeologists and in the McGlashan and Steiner collections include Dalton points, establishing that people lived in South Georgia during the Archaic period, and Clovis

Clovis points found in Georgia. National Museum of Natural History anthropology collections.

points, showing that people lived there in the Paleoindian period. At the Ocmulgee National Monument, archaeologists have found evidence suggesting that people lived in what is now South Georgia as far back as seventeen thousand years ago and perhaps longer. It boggles the mind to think that people lived contemporaneously with the great animals called megafauna that today we can only know through their fossils.

We are just beginning to discover the outlines of the ancient cultures that existed in South Georgia. We are not even sure how the people came to be there, and much remains to be learned from sites of early human occupation that now are underwater and date back to the last glacial age when sea level was several hundred feet below its present level.

The arrival of Europeans in the 1500s signaled the end of many indigenous cultures, which had thrived for thousands of years. The first contact between Native Americans and the expedition of Hernando de Soto in South Georgia in 1540 triggered an irreversible decline that occurred so quickly that little is left for us to see other than impressive mounds and artifacts that were resistant to deterioration by the elements.

military uniforms, weapons, battle flags (including the star-spangled banner that inspired our national anthem), medals, historical records, and paintings and photographs of battles. Other Smithsonian museums are engaged in this task as well, including the Smithsonian American Art Museum, the National Portrait Gallery, the National Museum of the American Indian, the National Museum of African American History and Culture, and the National Postal Museum. Given all of these collections, one might assume that finding artifacts about wars and South Georgia would be relatively easy, but such was not the case. Even though the land that constitutes South Georgia was claimed by competing forces—including Native Americans, the British, the Spanish, and the Confederate States of America—and was significantly impacted and shaped by the wars among them, few battles were actually fought there.

## The Revolutionary War

Prior to and during the Revolutionary War, Georgia had only one city that was considered a strategic asset—Savannah—and both sides of the conflict wanted to control it because of its status as a port and a seat of government. Savannah changed hands three times during the war, the last time in 1782 after the surrender of British troops at Yorktown in late 1781. Because it is located below the fall line, its story is part of my narrative.

Before the Revolutionary War, Savannah and the regions surrounding it were rife with division over the question of independence from Great Britain. The Loyalists were led by James Wright, who had served as governor of Georgia since 1760. The Patriots looked to people such as Button Gwinnett, George Walton, and Lyman Hall. Gwinnett, Walton, and Hall were transplants to Georgia who initially settled in Savannah but subsequently acquired property and built homes in the surrounding area. They were active in supporting the American cause and were recognized by their peers as leaders. The three men achieved lasting fame as the Georgia signers of the Declaration of Independence, which led to the first artifact in the Smithsonian collections that related to my search.

There is only one Declaration of Independence, and it is held in the National Archives. But copies of it were made that are of historical importance, among them the "Stone copies." In 1820 the printer William Stone was commissioned to create 200 copies of the declaration and its signatures. In 1823 one copy was distributed to each state, and two each went to the surviving signers of the declaration, including Thomas Jefferson, John

Adams, Charles Carroll, and other worthies. Stone did his work so well and his copies were so clear compared to the deteriorating original that his version is the one seen in today's textbooks. Some thirty of the Stone copies still exist, and according to values placed on them by auction houses, they can expect to be sold for $500,000 or more. But one copy will never be sold—the 201st copy, which Stone secretly made for himself. This copy was donated to the Smithsonian in 1888 by Stone's widow, Elizabeth. While there are no portraits or other objects related to South Georgians George Walton, Button Gwinnett, and Lyman Hall at the Smithsonian, their signatures are on Stone's copy of the Declaration of Independence, which now belongs to the American people.

All three Georgia signers played roles in Georgia's civic life, with Hall and Walton each serving as governor; Gwinnett ran but lost his race for the same position. Hall lived a long life and had many descendants, one of whom was his namesake, Lyman Hall, who became the second president of the Georgia Institute of Technology in 1896.

The Smithsonian collections commemorate the Revolutionary War with numerous artifacts, including George Washington's uniform and sword, but the cupboard is almost bare when it comes to South Georgia. The collections do contain a few artifacts related to the American heroes who participated in the battles for control of Savannah. One of them, General Nathanael Greene, who became known as the "Savior of the South," came to live on a plantation near Savannah after the war.

The stage was set for war in the state when Georgia Patriots took control of the city and port of Savannah in 1776. British government officials, including Royal Governor James Wright, decamped to England. Two years later, the British retook Savannah in the initial stages of what they called the "southern strategy," an effort to reinvigorate the campaign to defeat the Americans, which had stalled in the North. Wright returned to Savannah with the British troops after his two-year absence. Once the British had retaken Savannah and Charleston, troops began landing at the two ports and marching north under the command of General Charles Cornwallis to invade the Carolinas, their supply lines stretching back to the ports they had left behind.

As Cornwallis continued his advance into the Carolinas, George Washington asked his trusted associate General Nathanael Greene to take command of the southern forces and develop a battle plan. Greene, who was not a military man but was a student of military history, believed his army could only defeat Cornwallis by wearing him down using guerrilla tactics, particularly as the British supply lines became extended. Greene was right: by the spring of 1781 Cornwallis was forced north into Virginia, seeking help from the British fleet at Yorktown. Help did not arrive and, trapped by George Washington's forces, Cornwallis surrendered his army, triggering negotiations to end the Revolutionary War.

As peace negotiations dragged on into 1782, Savannah remained in the hands of the British. This situation did not sit well with Greene, who dispatched General "Mad" Anthony Wayne with a small contingent of troops to put pressure on the British in Savannah and keep them bottled up. Wayne's forces were a mixed bag of Georgia and South Carolina militia with a small core group of Continental soldiers. They were outnumbered in Savannah by the British forces and the indigenous people who were British allies. In spite of this disadvantage, Wayne managed to use his troops to good effect in small encounters and destroyed the supplies held at local plantations for the British troops. In July 1782 the British forces in Savannah surrendered the city and left, accompanied again by Wright, who this time departed Georgia for good. The surrender of Savannah solidified the Americans' negotiating position in the peace talks, ending any chance that the British would keep a southern foothold in the colonies.

The battles for Savannah are remembered in the Smithsonian collections through the American heroes who fought them. General Greene is recognized in the National Portrait Gallery with six portraits and a medal struck in his honor. In the National Postal Museum collections, two artifacts provide uniquely personal views of Greene. The first is a one-cent U.S. stamp issued in 1936 on which he is paired with George Washington, reflecting the admiration Greene held for Washington and the trust Washington placed in Greene.

One-cent George Washington and Nathanael Greene small die proof, postage stamp, 1936. National Postal Museum collections.

Free frank on Valley Forge cover, Nathanael Greene, April 2, 1778. National Postal Museum collections.

The second artifact is a letter, termed a "free frank on Valley Forge cover," which Greene wrote while he was with Washington at Valley Forge, serving as the quartermaster general of the Continental Army. The letter diplomatically pleads with Joseph Webb, a merchant in Connecticut, to provide desperately needed transportable baggage for the army as it prepared to leave the bleak conditions of Valley Forge. The letter is datelined "Camp Valley Forge, 2nd of April 1778":

> I request the Favour of you to procure as many good leather Portmanteaus, of about middling Size, as can be got ready to send or forward by the middle of May—The Number I shall expect from you will be about 200, at least. . . . I expect in a very short Time to remit you Money sufficient to pay for the Articles you may purchase. . . . I rely on your Zeal in the Publick Service to take upon you this Trouble, and to employ such Persons in the Business, as you shall think most likely to effect it to the best Advantage & with the greatest Dispatch.

It is no wonder that Washington saw Greene as one of his most loyal and talented generals. Greene stayed with him through the dark days of Valley Forge, proved to be an ingenious leader of armies, and was not above begging for luggage for the troops even when he was not sure he would have the money to pay for it.

General Anthony Wayne has three portraits and a medal struck in his honor in the National Portrait Gallery and a portrait in the Smithsonian American Art Museum. At the National Postal Museum, he is depicted on a two-cent U.S. stamp issued in 1929.

## After the Revolutionary War

Some artifacts look so ordinary, it is hard to believe they represent moments when history changed and hundreds of thousands of lives were affected. One such object housed in the Smithsonian collections is a patent model that dates to the period between the Revolutionary War and the War of 1812.

The story begins on a boat in 1792 with the meeting of two very different people. One was Catharine Greene, the widow of Nathanael Greene. In 1782 she and her husband had taken ownership of Mulberry Grove Plantation on the Savannah River, land that now belongs to the port of Savannah. The plantation was given to them by the state of Georgia for their services during the war. General Greene died suddenly in 1786, leaving Catharine with four children and the responsibility for managing Mulberry Grove. She was a determined woman who had spent the winter with her husband at Valley Forge and earned the admiration of George Washington and Alexander Hamilton. Washington visited Mulberry Grove twice to pay his respects to her.

Catharine was returning home on that boat in 1792, and at the time she was seeking a tutor for her children. She was introduced to a fellow passenger, a young man who had just graduated from Yale University. He had grown up on a farm in Massachusetts and evinced a special talent for making tools. His name was Eli Whitney, and he wanted to become an inventor. When he met Catharine Greene, Whitney was heading for a position on an estate in South Carolina, but he thought better of it and accepted Catharine's offer to tutor her children.

At Mulberry Grove, Phineas Miller, the overseer of the plantation and a fellow Yale graduate, introduced Whitney to the problem of removing seeds from short staple cotton. Short staple cotton was suited to the well-drained soil and regular rainfall of the South Georgia coast, but its bolls had many seeds attached to the cotton fibers. Removing the seeds by hand was the limiting factor in producing the large quantities of cotton needed to meet the market demand. With the encouragement of Miller and Greene, Whitney developed the idea for his cotton gin ("gin" being a shortened

Reproduction of Eli Whitney's cotton gin, donated by Eli Whitney. Treasures of American History at the National Museum of American History.

form of the word "engine"). By 1794 Whitney and Miller had sent a model to the Patent Office and patented the cotton gin.

One of the first copies of that patent model is in the collections of the National Museum of American History. I have used it, and it still works. In addition, an engraving of Eli Whitney's portrait, based on a painting made late in his life, is in the collections of the National Portrait Gallery. He appears to be saying, "I am smarter than you."

Whitney's cotton gin triggered a seismic shift in South Georgia, making the large-scale planting of cotton profitable and in turn generating a demand for laborers in numbers far beyond what had been needed for the rice plantations on the coast. The trafficking and enslaving of Africans in response to this demand and the scope of the slavery-based economy that resulted were principal causes of the Civil War.

Mulberry Grove Plantation was burned to the ground by General William T. Sherman's troops on their march to the sea during the Civil War. I visited the site in Port Wentworth just north of the Savannah airport, where a historical marker tells the plantation's story. The location where the house stood is protected on land owned by the port of Savannah, and the Mulberry Grove Foundation hopes to enhance the site for future generations.

## Two Men, Three Wars

Wars are decided upon and directed by nations, but it is individuals who fight them. Battles involve weapons and other artifacts that end up in collections, but it is often portraits that are the most expressive of the stories behind the conflicts. Such is the case for three wars that were fought between the Revolutionary War and the Civil War, cementing the fate of the indigenous people living in South Georgia and closing out the Spanish presence in Florida: the War of 1812, the First Seminole War (1816–1819), and the Second Seminole War (1835–1842). With the help of curators Wendy Wick Reaves and Ann Shumand of the National Portrait Gallery and Virginia Mecklenberg and Leslie Umberger of the Smithsonian American Art Museum, I found portraits of the Native American leaders whose lives spanned the era when the Creeks, whose ancestors had lived in South Georgia for thousands of years, lost all they had.

Two men in particular stand out. Although of different generations, they had much in common. Neither claimed South Georgia as home, but both spent time there. Both were born to Creek mothers and were fathered by white traders. Each had a lineage leading back to Creek leaders, but each chose a different path in the face of increasing numbers of white settlers seeking their land.

William McIntosh (Tustunnuggee Hutkee) was born in Georgia in 1775 and became an important leader of the Lower Creeks through his mother's ties to the Wind Clan and his own innate abilities. The lands he owned were principally north of the fall line, but his connections and influence extended into South Georgia. His choice was to collaborate with the whites and work within their system to secure a future for his family.

William "Billy" Powell, known today as Osceola, was born in 1804 in Alabama and then moved with his mother to Florida after the Upper Creeks lost the Battle of Horseshoe Bend in 1814. Like many other Creeks, he and his mother were assimilated into the Seminole Nation. As he grew older he watched as Creeks and Seminoles were forced to leave their lands in the face of broken treaties and ever-growing numbers of white settlers. In contrast to McIntosh, Osceola's choice was resistance, and his activities would take him across the Florida border into South Georgia as war was waged.

McIntosh came to manhood as President George Washington and then his successors sought to convince Native Americans it was in their best interest to adopt the norms of white culture—settle down on lands designated for them, cultivate farm crops, and establish schools and democratic governments. Washington appointed agents such as Benjamin Hawkins, who lived in Georgia on the Oconee River, to aid the Native Americans in their transition and to negotiate controversies that arose between them and white settlers. The Creeks appeared to fall in line, becoming the first tribe to be designated by the federal government as "civilized." During this time McIntosh met Hawkins and became adept at maneuvering the fine line between satisfying the U.S. government's expectations while maintaining his leadership position among the Creeks.

With the outbreak of the War of 1812, the Creeks and McIntosh had an unenviable choice. A splinter group of Creeks known as the Red Sticks had chosen to fight against the Americans in hopes of winning favorable terms from the British if they won the war. However, they made a critical error in slaughtering the occupants of Fort Mims, a lightly guarded frontier settlement in Alabama. Americans were outraged, and Andrew Jackson and his Tennessee militia were charged with defeating the Red Sticks. McIntosh and his Lower Creek warriors joined forces with Jackson. They earned a reputation for bravery and played a key role in the defeat of the Red Sticks in the Battle of Horseshoe Bend.

Jackson's hard-edged surrender terms in 1814 called for all Creeks, even those who had sided with him, to vacate twenty-two million acres of land in South Georgia and Alabama. Thus began the process that over the next forty years would lead to the forced evacuation of the Creeks from South Georgia to Oklahoma. McIntosh and his family were the exception. They were given special dispensations and land grants, and they were allowed to remain in Georgia. In 1816, in what became known as the First Seminole War, McIntosh once again decided that his family's interests were best served by siding with the Americans. This time, he and his forces joined Andrew Jackson to face the remnants of the Red Sticks, who were reinforced with people who had escaped slavery. Jackson, with McIntosh's help, won a decisive victory.

McIntosh was on the winning side of these battles, but he had engendered the enmity of many of the remaining Creek warriors, especially as the Creeks found themselves under continued pressure to surrender land to the growing numbers of white settlers. For reasons that have long been debated, McIntosh took the lead in signing the Treaty of Indian Springs in 1825, ceding the final parcel of Creek land in Georgia, which included the Ocmulgee Mounds, to the United States. It did not escape notice that McIntosh personally benefited financially from the treaty signing. Creek leaders who had not agreed to the treaty condemned him to death, citing the Creek law that land could not be negotiated without the full participation of the Creek people. An old antagonist, Menawa, who had led the Red Sticks at the Battle of Horseshoe Bend, was chosen to head the war party charged with killing him. Menawa did his work well, killing McIntosh and burning his plantation. In 1826 Menawa was a member of a delegation that went to Washington, D.C., to renegotiate the terms of the Treaty of Indian Springs. Although the U.S. government paid lip service to revising the treaty, nothing changed. Menawa and his family were forced to leave their home in Alabama in 1836 and were transported to Oklahoma. Records vary as to how and when he met his end, but it was far from the one he intended.

In 1825, while McIntosh was negotiating the Treaty of Indian Springs, Billy Powell, now known as Osceola, was emerging as a leader of the Seminoles. In 1835, feeling betrayed by the U.S. Army and the Indian agents, Osceola and his forces killed Indian agent Wiley Thompson and attacked U.S. Army troops in what came to be known as the Dade massacre. This attack triggered the Second Seminole War, and Osceola would become famous for his charisma and his ability to defeat and evade U.S. troops. During this conflict Osceola's Seminole allies used the Okefenokee Swamp as their base of operations, successfully hiding deep in the wilderness.

Osceola was captured by a ruse in 1837 when he came in under a white flag expecting to negotiate a ceasefire. Many Americans were incensed by this deceitful tactic. He was sent to Fort Moultrie, South Carolina, where he died of malaria in 1838. By the end of the Second Seminole War in 1842, most of the remaining Seminoles and Creeks were forced to go to Oklahoma. Osceola's ally Billy Bowlegs continued to live in Florida, but after leading the fight in the Third Seminole War (1855–1858), he and his family were forced to leave their home in 1858. He died in Oklahoma in 1859, the end of the line for the Creeks' claims to land in South Georgia.

The Smithsonian collections include portraits of the protagonists in this saga: McIntosh, Menawa, Bowlegs, and Osceola. McIntosh and Menawa were painted by Charles Bird King, and lithographs of their portraits were published by Thomas McKenney and James Hall in *History of the Indian Tribes of North America* (1836), a copy of which is held in the National Portrait Gallery. The portraits of McIntosh and his old adversary Menawa show men of stout character. Both are portrayed in clothing typical of the Creek people, but McIntosh also wears a vest and sash made of the tartan plaid of his Scots Irish father.

William McIntosh (Tustunnuggee Hutkee), lithograph of portrait by Charles Bird King. From Thomas McKenney and James Hall, *History of the Indian Tribes of North America* (1836). National Portrait Gallery collections.

Billy Bowlegs is represented by several portraits in the National Portrait Gallery collections, including a photograph made of him in 1858, just a year before he died. He is shown in his regalia, exhibiting a gaze that seems to reflect both his sorrow about the fate of his people and his pride at having resisted to the end.

The famous portraitist George Catlin painted Osceola after he was imprisoned in South Carolina and shortly before his death. Osceola was conscious of his appearance and for his portrait wore his best clothing; he was an indigenous man of unique character and status. His facial features, sometimes described as almost feminine, are softer than those of his fellow leaders. His eyes seem to see something at a distance, a sadness beyond our understanding. Catlin's painting of Osceola is held in the collections of the Smithsonian American Art Museum, while his lithograph of Osceola is in the National Portrait Gallery.

Portrait of Osceola (William Powell) by George Catlin, 1838. Smithsonian American Art Museum collections.

With the conclusion of the Seminole Wars, the presence of significant numbers of Native Americans in South Georgia, which had extended over thousands of years, became a thing of the past. Their language lives on in South Georgia in the lilting names of the rivers, such as the Altamaha, the Ocmulgee, the Oconee, the Alapaha, and the Willacoochee. McIntosh and Osceola have many places named for them. McIntosh Trail is a road McIntosh built between his lands in Alabama and Georgia. Counties, schools, towns, and one national forest are named for Osceola, and Florida State University uses his name for the mascot of its athletic teams. Some South Georgians believe that Ocilla, a neighboring city to my home of Douglas, was named for Osceola because he supposedly spent time in the area.

Today the Muscogee Nation calls Oklahoma its home. The Muscogee are closely related to the Poarch Band of Creek Indians, who live in southern Alabama. Both of these groups still look to the Ocmulgee National Monument as the ancestral home of their cultures, and this identity is now honored by the annual Ocmulgee Indian Celebration.

## The First Clough Arrives in South Georgia

The removal of the Creeks from South Georgia opened millions of acres to white settlers, attracting many from New England, where arable land had become scarce. To a young man living in Northfield, New Hampshire, the opportunity was timely, especially considering he was looking for a southern clime that would be kinder to his rheumatism. The Clough family had a long history in New Hampshire, having lived there since the 1720s and fought for American independence in the Revolutionary War and the War of 1812. But in 1837 John True Clough, my great-great-grandfather, left his home and set off to make a new one in Waycross, Georgia. He built a life as a farmer and public servant, and he married Serenia Catherine Miller. His stature in the white community increased when he joined the Georgia militia in the Second Seminole War, defending his fellow settlers against Seminoles based in the nearby Okefenokee Swamp. His service earned him the title of colonel, and he was rewarded with a grant of 490 acres of land.

John True Clough died in 1859 at the age of forty-one, having established a family that would survive the Civil War and go on to grow deep roots in South Georgia. The eldest of his five children, Jonathan Gilman Clough, my great-grandfather, enlisted as an eighteen-year-old in the Georgia cavalry in 1863 to fight for the Confederacy.

## The Civil War

While the Revolutionary War and the War of 1812 united Americans in the forging of a new nation, the Civil War was brutal and divisive. It left many families in mourning, in both the South and the North, including my own: cousins served in opposing armies and died for their respective causes. South Georgia as a place entered the Civil War story early on with the blockade of Savannah and later through two aspects that became infamous: Andersonville, the notorious Confederate prison, and Sherman's march to the sea.

South Georgia was far enough south to avoid the direct consequences of the Civil War until 1864, but when it arrived, life changed, not just until the end of the war but for decades to come. Wrapped up in this vortex of events were people, and among them were members of my family. It was not until I did the research for this book that I learned just what that entailed. When my great-great-grandfather migrated to South Georgia in 1837 from New

to conquer the city to neutralize it; a blockade of the port would serve just as well. It was not long in coming, but first Fort Pulaski had to be taken. Savannah is located a short distance up the Savannah River from the Atlantic Ocean, and Fort Pulaski was built in the 1830s along the river downstream from the city as a first line of defense.

The early phases of Fort Pulaski's construction were supervised by a young second lieutenant working on his first assignment after graduating from West Point, Robert E. Lee. Fort Pulaski entered Lee's life again in 1861 when Confederate president Jefferson Davis assigned him to review the coastal defenses of the Confederacy. Lee's role in building and later evaluating the defenses of Fort Pulaski is captured in a book in the collections of the Smithsonian Libraries: *Robert E. Lee and Fort Pulaski* by Rogers Young (1941). But even the great military mind of Robert E. Lee could not save Fort Pulaski. It was subdued by a weapon he had not seen in action—the rifled cannon—which punched holes in the eleven-foot-thick brick walls of the fort. On April 11, 1862, Fort Pulaski surrendered to Union forces, and the subsequent blockade helped cripple the South's economy during the war.

### *Andersonville Prison: Too Isolated for Union Troops to Reach*

About the time General William Tecumseh Sherman and his Union army arrived in North Georgia in the late fall of 1863, the Confederacy was wrestling with large numbers of Union prisoners, many captured in Virginia. The Confederate leaders were looking for a site where prisoners from different war sectors could be held while awaiting transfer to a suitable prison. The site needed rail access, but they also wanted it to be isolated enough to discourage raids by Union troops. The remote backwoods of South Georgia looked promising, and land that fit the bill was acquired near a small town called Andersonville in Sumter County, about sixty miles south of Macon. While prisons on both sides of the Civil War were known for their brutal conditions, the most notorious was Andersonville prison.

The facilities were still under construction when the first Union prisoners arrived in February 1864, but the captives were not sent on to other prisons as planned. Their numbers just kept growing, and by June the population had swelled to twenty-six thousand men, who were crowded into an area planned for ten thousand. The summer heat beat down, there was little to no cover, and their only drinking water was contaminated. Conditions in the camp became unbearable, and the rate of deaths increased alarmingly.

In July, on the advice of his camp surgeon, Captain Henry Wirz, the commanding officer at Andersonville, released five Union prisoners who were to deliver a signed petition to the federal government seeking a prisoner exchange to help reduce the inmate population. Wirz's offer was rebuffed by General Ulysses S. Grant with the support of President Abraham Lincoln. In his memoirs, an early copy of which is held in the Smithsonian Libraries, Grant explained his rationale: "Each man released on parole or otherwise, becomes an active soldier against us."

In spite of the conditions, the Confederates continued to send more prisoners, and the inmate population swelled to thirty-three thousand in August. Hundreds of men were dying each day of malnutrition and dysentery. Relief would soon come, but not in the form of humanitarian aid. As his forces were tightening the noose on Atlanta in July, Sherman authorized a cavalry unit to ride to Andersonville to liberate the Union prisoners. Although this effort failed, it raised fears that another attempt might be made, and in September Confederate forces began to move prisoners to other locations. From then on, the prison population at Andersonville never exceeded five thousand.

In May 1865, Henry Wirz was captured and became one of only a few Confederate soldiers to be executed for war crimes. He was hanged in November 1865 in Washington, D.C. A monument to honor Wirz was erected by the United Daughters of the Confederacy near the center of Andersonville.

When the toll was counted at the end of the war, nearly thirteen thousand Union prisoners had died in Andersonville prison. The dead were buried in mass graves during the war, but after the war a federal workforce, including the nurse Clara Barton, identified all but four hundred soldiers by using hospital records and letters containing identifying information, which were sent by relatives of the prisoners. Each soldier was given a headstone in the new Andersonville National Cemetery.

Of the thousands sent to Andersonville prison, five were members of the Clough family from New England who fought for the Union. Each was captured in a different battle, but all were sent to Andersonville. Two never left: Jedediah B. "Burchard" Clough and John D. Clough died in that terrible summer of 1864. On a beautiful sunny spring day in 2017, I made a pilgrimage to the cemetery with its thousands of white headstones gleaming in the sunlight. Because both Burchard and John Clough had enlisted in Vermont, their headstones were near each other: Burchard Clough, number 2694, and John Clough, number 3918. Standing there I felt a deep sadness for two

Headstone of John D. Clough, Andersonville National Cemetery. The inscription simply reads: "3918 / J. D. Clough / VT." He was a private in the First Vermont Heavy Artillery and was captured during the Battle of Weldon Railroad, Virginia, on June 23, 1864. My ancestor died on July 25, 1864, in Andersonville Prison.

young men who left small towns in New England to fight for the cause of the Union. I'm sure they thought they might die fighting, but not of disease in a miserable, hopeless prison in South Georgia. There was no Civil War glory here as depicted in my childhood textbooks. It was quite simply a horrible way for young men to die, and it did not have to happen.

Three of the Cloughs who spent time in Andersonville survived. One of these, Charles H. Clough, left desperately ill, and though he made it to the Union hospital in Marietta, Georgia, he died shortly afterward. He is buried in the Marietta National Cemetery. The other two, George W. Clough and John C. Clough, appear to have enjoyed long lives after the war. George Clough, who had lived in the same small town in Maine as Charles Clough, named his second son Charles.

Andersonville and its history are captured in a number of books in the Smithsonian Libraries, including the firsthand account of a survivor, John Ransom. Photographs of the monuments built by ten northern states in the Andersonville National Cemetery are in the collections of the Smithsonian

American Art Museum. Wirz can be found among the collections of the National Portrait Gallery in two photographs of his hanging taken by Alexander Gardner. One is titled *Adjusting the Rope* and the other *Springing the Trap*.

### *Sherman's March to the Sea*

In grade school history classes, I was told that General William T. Sherman was one of the greatest afflictions ever visited on Georgia. After my childhood I learned more about the complex story of the man who later in life became a member of the Board of Regents of the Smithsonian. On a visit to the military collections of the National Museum of American History, I was surprised to find myself standing before the campaign pin he personally designed for the march through Georgia. The inscription on the back of the pin reads, "Worn by Gen'l W. T. Sherman on the March to the Sea, 1864." I also saw the campaign hat and uniform he wore on the march. I was face to face with the man generations of Georgians have loved to hate.

The stage for Sherman's march was set in November 1863 in Chattanooga, Tennessee, when Union forces broke through the Confederate lines and pushed south, heading for Atlanta. By July 1864 Sherman was engaged in a battle for the city of Atlanta with Confederate general John Bell Hood. Among Hood's forces was a cavalry unit under the command of Joseph "Fighting Joe" Wheeler. Wheeler's ranks had recently been reinforced by calling up Georgia militia cavalry troops from South Georgia, including the Fourth Regiment, Company G, Georgia Volunteer Cavalry, which was known as the Wiregrass Fourth. Among the soldiers of the Wiregrass Fourth was my great-grandfather Jonathan Gilman Clough, who was nineteen at the time.

In late July, with Hood's army stymied, Sherman turned his attention to cutting off the railroad supply line that came into Atlanta from Macon to the south. After an initial strike by his cavalry, his infantry finished the job in August. Facing defeat, Hood attempted to demolish anything in Atlanta that might be of use to the Union forces—spiking cannons, destroying train engines, and blowing up train cars full of ammunition. After Atlanta fell to Union troops on September 2, 1864, Hood left for North Carolina, hoping Sherman would follow him. But Sherman had other plans.

Following his capture of Atlanta, Sherman set about destroying any infrastructure that might be of military or commercial value that Hood had left standing. The combination of battles and bombardments during the siege and the intentional destruction of its commercial and industrial

General William T. Sherman, albumen silver print, 1862, by unidentified photographer. National Portrait Gallery collections.

Pin made of buttons taken from the frock coat of General William T. Sherman, 1864. National Museum of American History Armed Forces History collections.

base left Atlanta devastated. Years later, Atlanta adopted the phoenix, a mythical bird that rose from its own ashes, as a symbol of its recovery from the Civil War.

While finishing in Atlanta, Sherman came up with an audacious plan and submitted it to President Lincoln and General Grant. According to Lloyd Lewis, on October 19, 1864, Sherman sent a telegram to Grant saying, "I can make the march and make Georgia howl." Splitting his army in two, he would send one division under General George H. Thomas to follow and destroy the remnants of Hood's army as he moved north. Sherman then would take the remaining sixty thousand men on a march from Atlanta to take Savannah. This army would take little in the way of provisions, allowing it to proceed with dispatch while foraging for what it needed from the countryside along the way. It would burn the fields and plantations of wealthy Georgians and wreck any infrastructure that might be used by the Confederate army, such as railroads and bridges. Sherman even planned to cut telegraph lines to keep the Confederates from finding out where he was going, with the result of cutting off his own communications. His would be a stealth force, intended to confuse troops that might have some ability to attack and to strike terror in the local populace. Despite some trepidation, Lincoln and Grant agreed.

On November 15, 1864, Sherman set out from Atlanta, traveling on a course that was essentially a straight line to Savannah. His forces were divided into left and right units that could move apart and come together as needed. According to Sherman's orders, his troops were not to destroy towns unless they resisted. The test case would be Griswoldville, which Sherman's forces reached on November 22. It was Griswoldville that sent me on another excursion into the Smithsonian collections.

The town was founded by Samuel Griswold, who moved to Georgia in 1820 from the center of the weapons production area of Connecticut. His primary source of income before the Civil War was the manufacture of cotton gins, but when the cotton trade was stymied by the Union blockade of the southern ports, he switched back to his family's stock-in-trade, weapons. Georgia governor Joe Brown provided Griswold's first break when in a moment of inspiration he asked Griswold to produce pikes, a type of spear that became known as "Joe Brown pikes." Brown's enthusiasm for pikes exceeded his logic, since the handheld weapons of this war were rifles and pistols, which were steadily becoming more deadly. There is no record that any Joe Brown pikes were used successfully in battle, but several are included in the weapons collection of the National Museum of American History.

Griswold and Gunnison revolver, 1862. National Museum of American History Armed Forces History collections.

Griswold was more successful with a handgun that was a knockoff of the famous Colt Navy revolver. It became known as the Griswold and Gunnison revolver and was distinctive because its frame was brass instead of steel, which was hard to come by in Georgia during the war. Only about thirty-five hundred of these revolvers were made, and they are collector's items today. The long-suffering protagonist of the popular television series *Hell on Wheels*, Cullen Bohannon, is a former Confederate officer who wears a Griswold and Gunnison revolver. One in good condition is found alongside Joe Brown's pikes in the Smithsonian weapons collection.

Griswoldville's strategic value to the Confederacy made it worth defending, and the first Confederate unit to arrive was General Joe Wheeler's cavalry. Wheeler understood he was outnumbered and faced a battle-tested infantry force. After a series of attacks, Wheeler left the field, hoping to head off the leading edge of Sherman's army, which was moving toward Savannah. As Wheeler was leaving, a unit of the Georgia militia arrived, consisting mainly of boys and old men. The militia attacked the Union forces multiple times to no avail, resulting in 51 killed, 471 wounded, and more than 600 captured. Union casualties were minimal, and the federal army proceeded to sack and burn Griswold's factories and warehouses.

Meeting little further resistance along their way, Sherman's forces arrived at the outskirts of Savannah on December 10, having left Atlanta less than a month before. The ten thousand Confederate troops in and around Savannah under the command of General William Hardee made an initial show of resistance, then lingered for a few days before deciding to leave

Jonathan Gilman Clough, great-grandfather of the author, served as a corporal in the Wiregrass Fourth, Georgia Cavalry.

rather than fight. Their departure spared Savannah the fate of Atlanta. Sherman met with the mayor of Savannah, who promised to hand over the city to him. Sherman realized that an intact Savannah with thousands of cotton bales sitting on its wharves was worth more to him than a Savannah destroyed, so he followed his tendency to be lenient with cities or towns that did not oppose his forces. On December 21, 1864, Sherman sent a telegram informing President Abraham Lincoln that Savannah was his Christmas present.

The success of Sherman's march to the sea inspired a catchy tune called "Marching through Georgia." The Smithsonian Center for Folklife and Cultural Heritage has a recording in its collections sung by Ernest Stoneman. Predictably, northerners loved the song and southerners hated it. Eventually Sherman himself came to dislike it because of the negative emotions it stirred up after the war was over.

After the fall of Savannah, the war for all intents and purposes was over. A few months later on April 9, 1865, Robert E. Lee surrendered his forces to General Grant at Appomattox. On April 26 Joseph Johnston surrendered the Confederate forces under his command to General Sherman. Among those troops was a corporal in the Confederate cavalry, Jonathan Gilman Clough, who luckily for me had survived the war.

### *The Day the Confederacy Ended: In a Pine Barren in South Georgia*

I was standing in front of a collection cabinet in the National Museum of American History with curator Harry Rubenstein, looking at two elegant engagement rings that had once belonged to a young woman named Varina Howell. In 1844 when she was seventeen, a thirty-five-year-old Mississippi plantation owner proposed to Howell, and she accepted. The plantation owner was Jefferson Davis, whose first wife had died a decade earlier. In 1845 Varina Howell, a vivacious, intelligent, independent woman, married Davis despite her parents' reservations. Davis, a graduate of West Point, went on to serve in the Mexican-American War, as the U.S. secretary of war, and as a U.S. senator from Mississippi. During his tenure in the Senate he was named a regent of the Smithsonian, a post he held for three years. In 1861 he made the fateful decision to become the first, and only, president of the Confederate States of America.

On April 3, 1865, Jefferson and Varina Davis fled Richmond shortly before the city fell to Grant's forces. They headed south with an entourage, stopping for short stays at different places along the way. Davis was hoping somehow to get to Texas, where he expected to find sympathetic troops who could serve as a rallying point for the Confederate cause. On April 15, 1865, Abraham Lincoln was assassinated, and many presumed the president of the Confederacy was complicit in the deed. Interest in Davis's whereabouts became keen, especially as a $100,000 reward was offered for his capture.

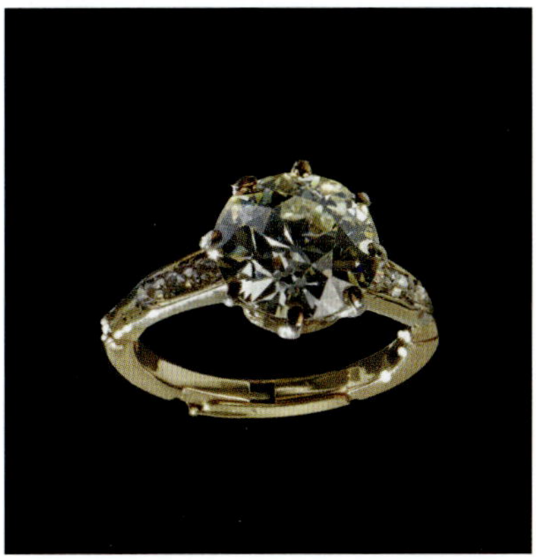

Engagement rings of Varina Howell Davis, wife of Jefferson Davis, president of the Confederate States of America. National Museum of American History collections.

By May 9, Davis, his wife, an enslaved person who had been with his family for years, and a few soldiers had made their way to the vicinity of the small South Georgia town of Irwinville—a hundred miles south of Macon and about thirty-five miles west of my home of Douglas. The party camped in a pleasant pine barren, thinking this isolated spot would be safe for a day or two. Unbeknown to them and to each other, two units of Union soldiers who were searching for Davis also arrived that same night on different sides of the camp. In the morning, a ground fog obscured visibility. The Davis camp was just coming to life, and Varina Davis had placed her shawl around her husband to ward off the chill when one of the Union units heard a noise and fired. The shots missed Davis and his companions and instead hit the Union soldiers on the other side. Davis surrendered, but two Union soldiers died of friendly fire.

Today the capture site features a memorial placed by the state of Georgia. On the way there you pass through Irwinville where a sign welcomes you, noting it is home to the Irwinville Farmers basketball team, the state's Class C high school champions back in the late 1940s and early 1950s. The main businesses are facilities for peanut processing and a cooperative cotton gin. The memorial is located a few miles from the town and is easy to miss. When I visited it in 2016, no one was there when I arrived, and no cars passed on the road for the hour or so of my visit. The museum, which is only open one day a week, was closed when I was there. It was built by the Works Progress Administration in 1939 and is a little worn with age, but it still has a pleasant appearance.

The grounds feature two attractions. The first is a stone memorial with a bust of Jefferson Davis; his name is on the front. The back states, "ON THIS SPOT / MAY 10, 1865 / PRESIDENT JEFFERSON DAVIS / WAS MADE A / PRISONER OF WAR / BY FEDERAL TROOPS." This monument appears in photographs in the collections of the Smithsonian American Art Museum. Erected in 1935 by the United Daughters of the Confederacy with help from the state of Georgia, it was dedicated by Governor Eugene Talmadge. As you pass along a shaded path cut through the woods, there is a second memorial: a metal plaque announcing, "KILLED ON THIS SPOT / May 10, 1865 / JOHN RUPERT, COMPANY C, 4TH MICHIGAN CAVALRY FEDERAL ARMY . . . / JOHN HINES CORPORAL COMPANY E, 4TH MICHIGAN CAVALRY FEDERAL ARMY . . . / THEY DIED BRAVELY IN THE LINE OF DUTY, MISTAKENLY SHOT BY FRIENDLY / WISCONSIN CAVALRYMEN." The plaque tells us that Rupert was twenty-one and Hines eighteen when the two enlisted in the war and that both of their bodies were moved to the national cemetery at Andersonville.

The site was hushed with only the sound of breezes whispering through the longleaf pines. I thought about the two young Union soldiers who had survived the war and were probably looking forward to the thrill and importance of capturing Jefferson Davis. And then they were shot dead by their own troops. I imagined Varina and Jefferson Davis, who surely never thought they would be camping out near Irwinville, Georgia, while running from Union troops. Even as they got up that morning, they probably thought they might escape to lead a comeback by the Confederacy. But a South Georgia pine barren was the end of the line for it all—for the Confederacy, for the war that had brought so many deaths, and for Jefferson Davis.

The Smithsonian collections contain many portraits and sculptures of Davis, and as president of the Confederacy, he appears on Confederate coins, bills, and stamps. As a historic figure he is also on a U.S. stamp held by the National Postal Museum. Varina Davis is shown in several photographs and a miniature in the National Portrait Gallery. She lived on long after her husband's death and became a correspondent for a New York newspaper.

## A Sea Change in South Georgia

The period from the Revolutionary War through the Civil War spanned only ninety years, but it involved six wars, massive population shifts, and the expansion of a young nation's boundaries south and west. At the beginning of this era, South Georgia belonged to the Creeks, but by the time the Second Seminole War ended in 1842, the Creeks' presence was essentially eliminated. In their place came thousands of European settlers along with migrants from other locales in the United States, like my ancestor John True Clough, who came from New Hampshire in 1837. The population increase in South Georgia during this time also included large numbers of enslaved people brought from Africa to work the cotton fields and rice plantations. In the midst of all this upheaval, the Spanish era in Florida came to an end, along with its influence on the boundary regions where Georgia meets Florida. And Georgia, after briefly seceding from the United States of America to join the Confederate States of America, ended up after the Civil War where it started, once again a part of the United States of America.

The Civil War left Georgia in shambles. Thousands of the young men sent to battle never came back, leaving a demographic hole. Fortunately for me, my great-great-grandfather Cuyler Hilliard and my great-grandfather

Jonathan Gilman Clough, both of whom fought for the South, survived. But the place they knew before the Civil War was gone. The railroads—lifelines for transporting crops and goods—were destroyed by Sherman's forces. Black people, who made up 45 percent of the population of the state, were freed but thrown adrift. Lincoln's vision for reconciliation was lost with his assassination. After Reconstruction, as southern whites regained power, they passed Jim Crow laws that created an artificial, segregated society that was preordained to fail.

For decades following the Civil War, educational opportunities in the South were sparse for whites and worse for blacks. Even the 1940s of my childhood were tinged with the long legacy of the Civil War. When the crops were in, my cousins and I would ride horses out into the countryside. We would come upon tenant farmers, both black and white, whom I could barely understand. It was as if they spoke another language, and the vacant eyes of the children foretold their future.

Artifacts in the Smithsonian collections about South Georgia during this period are meager, but the stories are profound: a secret copy of the Declaration of Independence with the signatures of South Georgians Button Gwinnett, George Walton, and Lyman Hall; Eli Whitney's patent model for the cotton gin; portraits of William McIntosh and Osceola, tragic figures in the removal of the Creeks from South Georgia; the business card of a firm advertising the buying and selling of people who were treated as property; photographs of the monuments to the Union dead at Andersonville prison and of the commandant, Henry Wirz, at the moment of his death by hanging; the campaign badge designed and worn by General William Tecumseh Sherman on his march through Georgia; and the engagement rings of Varina Davis, wife of Jefferson Davis, who was with him the day he was captured in Georgia.

If there is an aura of tragedy around many of these objects, it is rightfully so. This history began on a hopeful note with the victory over the British in the Revolutionary War, but with the growth of the institution of slavery, the eradication of the Creeks, and the bitter ordeal of the Civil War, South Georgia became a place wanting in optimism and enlightenment. In spite of it all, people would endure. Both blacks and whites lived on with hope for better times, regardless of the long odds they faced. Among them were members of my family.

CHAPTER TWELVE

# Gullah Geechee

## *A South Georgia Culture unto Itself*

My search for artifacts about South Georgia had taken me to almost every collecting unit at the Smithsonian, but the biggest surprise still awaited in one of the smallest collections. Early in my tenure as secretary of the Smithsonian I worked my way through visits to all of the museums and research centers, concentrating on the big ones first and watching as the smaller ones moved up on the not-yet-visited list. Among these was the Anacostia Community Museum, established in 1967 as a model for a museum that serves its local community. The Anacostia neighborhood is located southeast of downtown Washington, D.C., south of the Anacostia River, and its population is primarily African American.

From my first visit I was impressed by this small museum and its unique role. It has strong community support and provides educational opportunities for local schools that are unable to fund field trips to the National Mall. When I asked the director at that time, Camille Akeju, about upcoming exhibitions, I was told about an ambitious project that would focus on the Gullah Geechee culture of the Low Country along the Georgia and South Carolina coasts. It would feature the work of Dr. Lorenzo Dow Turner, the man who discovered the African roots of the Gullah language. This project struck a chord with me, given my own background in South Georgia. What I did not fully appreciate at the time was that the collections of this museum contained some of Dr. Turner's papers related to places I had visited before. This was another of the wonderful surprises that popped up during my search through the Smithsonian collections.

## Gullah Geechee Culture

William T. Sherman famously said "war is hell," and after his march to the sea, most Georgians agreed. Yet in the midst of the ransacked countryside and twisted railroad tracks, one group of South Georgians simply got left behind—a different sort of legacy from the Civil War. They had been brought to colonial America from Africa to work as slaves on the Low Country rice plantations like the Woodmanston Plantation of the LeConte family. These plantations were often in isolated locations and insulated from society by waterways, swamps, and wetlands. There, enslaved people incubated a unique culture that combined the traditions and foodways of their homelands with aspects of the new life they were forced to accept.

Because the economics of rice plantations did not work without slave labor, the adoption of the Thirteenth Amendment to the Constitution in 1865 signaled the end of this way of life. Many of the grand homes were torched by Sherman's troops, their remaining chimneys silhouetted against the sky as reminders of what was and would never be again. White former slaveowners took what they could and left. With nowhere to go, the formerly enslaved people remained behind and made homes on the offshore islands and on the banks of rivers along the coast. The majority of them had come from the Windward Coast of Africa, principally from today's Liberia, Sierra Leone, and Guinea—places where growing rice was an ancient skill. Their languages, religions, songs, stories, and ability to make baskets from grass gave them much in common. Living in the Low Country for more than 150 years before the Civil War and sharing a common heritage, they created their own language, which today we call Gullah.

The people themselves became known as Geechee or Gullah, the former name applied to those living on the islands off the Georgia coast and the latter to those living on the islands off the South Carolina coast. The etymology of the names "Geechee" and "Gullah" is not clear. Historian Wilbur Cross suggests that they derive from comparable names on the Windward Coast, while others believe they derive from names in the southern United States. For example, "Geechee" might be a derivative of the name of the Ogeechee River near Savannah, which was given to the river by the Creeks.

As a child, I had been captivated by Joel Chandler Harris's tales of Uncle Remus, especially stories of the antics of Brer Rabbit. Since I was prone to making trouble myself, Brer Rabbit became something of a role model. He taught me that even though you might not be the smartest, fastest, or strongest, you can win most of the time by being clever, especially if you add a dash of humor.

Although I did not know it, Uncle Remus spoke in a dialect that Harris adapted from Gullah, and the stories were variations on Gullah Geechee folktales. In 1862 at age sixteen, he had apprenticed as a printing compositor on the Turnwold Plantation in Putnam County, Georgia, where he worked with the owner, William Addison Turner, to publish a newspaper, the *Countryman*. Before the plantation was nearly destroyed by Sherman's troops on their march to the sea, Harris learned the printing trade and spent time with the enslaved people, learning their folktales and stories. This experience, combined with his storytelling ability and sense of humor, led to the creation of the character Uncle Remus and his magical world. For most of his career Harris divided his efforts between printing newspapers and his writing. Finally, his stories proved to be so popular that he retired from printing and went to writing full time. Harris is represented in the Smithsonian collections in two portraits in the National Portrait Gallery and by many books in the Smithsonian Libraries.

As an adult, I enjoyed Pat Conroy's autobiographical novel, *The Water Is Wide*, about Conroy's two years of teaching Gullah children on Daufuskie Island on the border between Georgia and South Carolina. I later encountered the Gullah Geechee culture in person when a study I was conducting to reassess the magnitude of the 1886 Charleston earthquake took me to Edisto Island. But it was in the collections of the Anacostia Community Museum of the Smithsonian that I "met" Dr. Lorenzo Dow Turner.

## Lorenzo Dow Turner

In 1901 the Reverend Charles Albert Tindley wrote an African American spiritual titled "I'll Overcome Someday," which begins with the words "In my heart I do not yield, / I'll overcome someday." Some have argued it is a predecessor of the famous civil rights anthem "We Shall Overcome." Tindley's words epitomize Turner who, despite the challenges he faced, conducted research to understand the Gullah language and its origins. Many people, including linguists, contended that Gullah was a second-class dialect invented by people unable to learn English. Turner was told he was wasting his time.

Turner was born in 1890 in Elizabeth City, North Carolina, to an African American family whose ancestors had been enslaved. His family was middle class and well educated. His father, Rooks Turner, earned a college degree and established a good life for his family in Elizabeth City, only to see it turn bad because of the racism of the time. The family moved to Maryland, where Lorenzo Dow Turner earned a bachelor's degree at

Lorenzo Dow Turner with the device he used to record his Gullah Geechee "informants." Photographer unknown. Anacostia Community Museum collections.

Howard University in 1914; he earned a master's degree at Harvard in 1917. After a teaching stint, he was admitted to the University of Chicago and received his PhD in English literature in 1926. He tried his hand at publishing a newspaper, but by 1929 it had folded. The failure of his newspaper turned out to be one of the best things that ever happened to him, since it led to his taking a summer teaching position at South Carolina State College at Orangeburg. There, he first heard students speaking in a dialect he could not understand, which set him on a course for the remainder of his life to identify the roots and nature of the Gullah language.

Turner decided he needed formal training in linguistics, and in 1931 he became the first African American member of the Linguistic Society of America. With little guidance, he set out to document and record the language he had first heard at Orangeburg. In unbearable heat and humidity, he lugged around a hundred-pound recording device that inscribed sounds on an aluminum disk. He recorded speakers he called "Gullah informants" on the sea islands of Georgia and South Carolina, and they were thrilled when he played their voices back to them. He also documented their music and folktales and took photographs of the people and their homes.

In 1932 Turner was asked to deliver a lecture at Yale on his early findings, and those present were impressed. But he knew he could not yet say with certainty how the Gullah language came to be. Convinced that he needed to know more about the African languages of the Windward Coast, he obtained support in 1936 to undertake studies in the African languages program of the University of London. There, he began to see clearly the similarities of Gullah phrases and words to those found in African languages.

During this time he was invited to set up a tent at the 1937 Exposition Internationale in Paris to record visitors from Africa, particularly those from the Windward Coast. He realized on listenening to the Africans that he was hearing languages similar to but not the same as Gullah. When he played his Gullah recordings for the Africans, they said they understood some of the words they were hearing. This experience confirmed the connection Turner was seeking: Gullah was a creole language created by people who fused their native tongues with what they needed from the language of their new home.

After Paris, Turner traveled to Brazil in 1940 to study the languages used by former enslaved people there, some of whose ancestors came from the Windward Coast and Central and West Africa. This work solidified his case for the African origins of Gullah and expanded understanding of how slavery had created a multinational diaspora.

Back home from his travels, Turner began work on a seminal book titled *Africanisms in the Gullah Dialect*, which was published in 1949. Early editions of this book are in the collections of the Anacostia Museum and the Smithsonian Libraries. Today Turner's book and his research are accepted as having founded the basic concept for understanding creole languages. Turner passed away in 1972 after a life spent helping others understand a people and a culture that could have been lost.

### *The Turner Collections*

Turner meticulously documented his research with hundreds of photographs, recordings, journals, artifacts, and writings. Preserving his work generated interest not only from linguists but also from musicologists, art historians, and historians in general. He worked at several institutions, and the archives at Indiana University, Northwestern University, the University of Chicago, and Fisk University hold portions of his papers and artifacts. In 2003 his widow, Lois Turner Williams, donated the items she possessed related to her husband to the Smithsonian and in particular to its Anacostia Community Museum.

While working on this book, I met with curators Portia James and Alcione Amos of the Anacostia Museum to view items from the Turner collections, particularly those that related to *Word, Shout, Song: Lorenzo Dow Turner Connecting Communities through Language*. This exhibition was on view from August 2010 to July 2011 and attracted a record number of visitors to the museum. I was one of the visitors, and it was a memorable experience, especially given my interest in the Gullah Geechee people of the South Georgia coast.

As I leafed through Turner's photographs, I was struck by the images from real-life settings, not studios. Turner made recordings in South Georgia coastal communities such as Harris Neck, St. Simons Island, and the Sapelo Island towns of Hog Hammock and Raccoon Bluff—communities that were still isolated, sometimes for lack of a bridge to the mainland.

A ring shout. Photograph by Lorenzo Dow Turner. Anacostia Community Museum collections.

The original aluminum disk used by Lorenzo Dow Turner to record the Mende funeral song sung by Amelia Dawley. Anacostia Community Museum collections. Curator Portia James is with the author.

His photographs captured the dignity and spirit of people who had lived a hard life but had lived well. His photograph of a ring shout performed by a group of Gullah Geechee men and women became iconic, documenting a style of singing and dancing that he would find was rooted in African traditions.

Some of the many recordings Turner made were of people singing, and one was of Amelia Dawley of Harris Neck singing a song she said was used for funerals. She had learned it from her mother, Tawba, who had learned it from her mother, Catherine, both of whom had been enslaved. During my visit to the Turner collections, I heard the Dawley performance as it was recorded by Turner. I sat transfixed, listening to her sing as I followed along using a translation written by Turner on paper that now was yellowed with age. What an amazing experience.

Scholars following in Turner's footprints have shown that Dawley's song is associated with the Mende people of Sierra Leone. There, women sing it as the deceased's body is being prepared for burial and as they make dishes of rice, which become an offering for the dead. Here is a literal translation of the song:

> Come quickly, let us work hard, the grave is not yet finished; his heart is not yet perfectly cool [at peace].
> Come quickly, let us work hard, the grave is not yet finished; let his heart be cool at once.
> Sudden death cuts down the trees, borrows them, the remains disappear slowly.
> Sudden death cuts down the trees; let it be satisfied, let it be satisfied at once.
> Sudden death cuts down the trees, borrows them, a voice speaks from afar.

The Turner collections at the Anacostia Community Museum also include a large number of objects and photographs from his work in South America and Africa. They display the extent of his scholarship and his remarkable commitment to understanding the entirety of the Gullah Geechee story.

## The Human Connection

As the *Word, Shout, Song* exhibition explained, Turner's recording of Amelia Dawley singing the Mende funeral song proved to be the connection that linked her Geechee family in South Georgia with their roots in Sierra Leone. Making that connection was the anthropologist Joseph Opala, who became interested in the slave trade when the Peace Corps sent him to Sierra Leone in 1974. He knew that thousands of Africans from Sierra Leone had been captured and sent to Georgia and South Carolina.

Back in the United States, Opala turned his attention to the descendants of the people who had been brought to Georgia and South Carolina from Sierra Leone. As he worked with the Penn Center on St. Helena Island, South Carolina, and the institutions that held Turner collections, Opala realized that the Turner recording of Dawley could be the link he was looking for. He set out to find the family of Amelia Dawley and met her daughter, Mary Moran, who still lived in Harris Neck. Moran had been taught the Mende funeral song by her mother.

Returning to Sierre Leone, Opala and his colleagues found that many of the old traditions and songs had been lost with the coming of Christianity and Islam. However, in the Mende village of Senehun Ngola they discovered a woman named Baindu Jabati whose grandmother had taught her the original Mende version of the funeral song. So in 1997, Opala and his colleagues arranged a "Moran family homecoming" in Sierra Leone. Mary Moran from Harris Neck, Georgia, and Baindu Jabati of Senehun Ngola sang the funeral song together. The song had made the round trip back to Africa.

## Living Musical Traditions

As Lorenzo Dow Turner discovered, music is one of the living traditions of the Gullah Geechee culture. In Africa songs and music are part of work, celebrations, and religious ceremonies. When Africans were brought to the Low Country, their traditional use of drums was outlawed because slave-owners believed they could be used as a means of communication between people on different plantations. With no drums, Africans in America resorted to rhythmic clapping, stomping on the floor, and striking the floor with sticks. At the same time, enslaved people began to weave concepts of Christianity and its hymns into their music.

A common motif in Gullah Geechee music is call-and-response, in which someone sings a phrase and others reply. Call-and-response songs in which enslaved people joined their voices provided a sense of group solidarity and rhythm that helped them endure working in the hot sun.

By the late 1800s, after slavery had ended, the number of people who knew the old songs and singing styles had begun to diminish, and traditional Gullah Geechee music was at risk of disappearing. Groups like the Georgia Sea Island Singers, formed in the early 1900s, have helped continue these music traditions. The Sea Island Singers, including Bessie Jones in her time and Douglas and Frankie Quimby who followed, have kept Gullah Geechee music alive through performances and a commitment to teach the music to new generations.

Along with other forms of Gullah Geechee music, the ring shout was also dying out until it was recorded and photographed by Lorenzo Dow Turner in the early 1930s. In this form of music, the shout is not a vocal outcry but a rhythmic movement around a sacred object. As Turner surmised, the word derives from the Arabic word *sha'wt*, which refers to the counterclockwise rotation of pilgrims around the Kaaba in Mecca. The ring shout involves a slow, halting circular rotation synchronized to the rhythm of a stick hitting the floor during a chant or a call-and-response song. Smithsonian Folkways has recordings of ring shouts performed by the famous McIntosh County Shouters of the Georgia coast and offers them to the public. Today descendants of the original Shouters continue to keep the tradition alive with public performances.

Alan Lomax and Moses Asch, who were dedicated to documenting our nation's folk music, made recordings in the 1950s and 1960s of the music of the Gullah Geechee culture. The Smithsonian Center for Folklife and Cultural Heritage acquired the collections of Moses Asch in 1987 and incorporated them into the Smithsonian Folkways collections. This archival resource contains many of the early recordings as well as current titles from the Sea Island Singers. Folkways also provides guides for teachers to help students learn about Gullah Geechee culture and music.

In 2013 I received an email from a friend with a photograph taken in the South Georgia coastal town of Darien, population two thousand, in McIntosh County. The photo shows a large banner, displayed on the front of a business that had seen better days, proclaiming, "The Smithsonian Is Coming." Actually, the Smithsonian was not about to open a museum in an abandoned store in Darien, Georgia, but the Smithsonian Institution Traveling Exhibition Service was on its way. This exhibition was on a circuit throughout Georgia in partnership with the Georgia Humanities Council and was titled *Museum on Main Street—New Harmonies: Celebrating American Roots Music*. The core of the exhibition was designed by the Smithsonian, but it offered the opportunity for participation by local music groups. In the case of Darien, the McIntosh County Shouters were featured. I was told this exhibition treated hundreds of people to performances of Gullah Geechee music.

I am proud that the Smithsonian has helped preserve the work of Dr. Turner in the collections of the Anacostia Community Museum and in the recordings of Smithsonian Folkways. As future generations learn about the traditions and lives of people who created one of America's unique cultures, I hope they also come to appreciate the role played by a place called South Georgia in its development.

CHAPTER THIRTEEN

# South Georgia Arts and Artists

My search to find South Georgia in the Smithsonian collections led me to crawfish, cow killers, snakes, trees, birds, giant sloths, Clovis points, Civil War weapons, and a wide assortment of other specimens and artifacts. But in those cases the Smithsonian art collections had only served in a secondary role as background for other subjects. The art collections had not yet been examined to see what insights they might offer in themselves that would help me understand my home.

Art by definition includes literature, visual art, and music, and the Smithsonian collections include works in all of these categories. However, given that the collections of the Library of Congress, the National Archives, and the National Gallery of Art also cover these topics, the Smithsonian long ago concluded that its collections in these areas should serve specific mission-related roles. For example, the Smithsonian Libraries collections excel in their coverage of natural history but don't focus on literature for literature's sake. The Smithsonian music collections are among the best in the world for folk or roots music but not for classical music. When it comes to art, the Smithsonian has no single comprehensive collection but rather a number of first-rate smaller collections of, for example, American art, contemporary art, or portraiture.

While the Smithsonian art collections are not as comprehensive as those of natural history, experience had shown me it was not the quantity but the quality and the element of surprise that made the search for something about my home worthwhile. I was ready to take the plunge.

## The Arts in South Georgia: Birthed by a Long Labor

If Willie Nelson is right that art comes out of poverty and hard times, Georgia—and South Georgia in particular—provided plenty of both in the decades following the Civil War. The first sign that some change might be possible came in the 1870s when journalist Henry Grady, who was to become the managing editor of the *Atlanta Constitution*, coined the term "New South." This expression embodied Grady's vision to broaden the base of the region's economy from almost total dependence on agriculture to incorporate more industry, and to improve education for the citizens of Georgia. Grady and others subsequently put their shoulders to the wheel in an effort to make the New South more than a slogan. Grady died in 1889, six years before the 1895 Cotton States and International Exposition in Atlanta, at which six states of the old Confederacy pronounced to the nation that the New South had arrived and a new era had begun. At the exhibition, the states displayed wares demonstrating their progress and touted their new commitment to education. Also included were a Women's Building, a Negro Building, and a Government Building, each offering optimistic views of the future.

In the Government Building, the Smithsonian was prominently featured in an exhibition curated by Assistant Secretary G. Brown Goode. Goode's report to the Smithsonian Board of Regents began by saying, "The exhibit made by the Institution was not as satisfactory as it was planned to be owing to the small amount of money allotted for its preparation." But he concluded, "Although small, [the exhibition] proved itself to be both attractive and instructive." In spite of Goode's complaint about funding, the exhibition included displays on just about everything the Smithsonian collections could provide: mammals, insects, birds, fishes, minerals, and types of humans. It even had a display on the origin of games. The intent was to demonstrate how the Smithsonian and its scientists and collections were expanding our understanding of life. The regents expressed their appreciation to Goode for his success in helping the citizens of the South get a glimpse of what a great museum could provide.

The Cotton States Exposition ended on a high note, but its promises were founded more on hope than reality, as the economies of the southern states continued to lag those of the northern states for decades to come. Nevertheless, in the twentieth century, doors began to open for creative expression. Artists began to emerge who used literature, art, and music to speak to both the beauty and the heartaches of life in South Georgia, creating something of a renaissance. And, I was to learn, the Smithsonian had a hand in documenting that revival of the arts, which occurred as South Georgia began to pull itself out of the morass created by the Civil War.

## Writers and Literature

My search for South Georgia writers did not begin in the Smithsonian Libraries, and writers were not even on my mind at the time. In an effort to identify prominent South Georgians in the collections, I had asked the curators at the National Portrait Gallery if there was a way to sort out people with South Georgia roots. Of course, I knew that Jimmy Carter, born and still living in Plains, would be there because a portrait of each president hangs in the Hall of Presidents. I was thinking more in terms of less well-known people who might have earned a place in the collections.

I was fortunate that curators Wendy Wick Reaves and Ann Shumard were enthusiastic about helping, and they invited me to see what they had found. Fortunately, most of the twenty-two thousand portraits in the collections have been digitized, and place of birth is one of the search terms for the database. Reaves and Shumard had quickly identified all the people who were born in Georgia. Then, using a map of the state, they winnowed the candidates down to those born in places located below the fall line. By the time I arrived, they had arrayed the final results for my review.

There were more people than I expected, and the majority had been politicians, war heroes, musicians, or athletes. However, among the images I was pleasantly surprised to see a photograph of Sidney Lanier, Georgia's most famous poet. The photograph was taken in 1875, only six years before Lanier died at the age of thirty-nine from tuberculosis, contracted when he was held prisoner during the Civil War by Union troops. Born in Macon in 1842, Lanier was multitalented, both an accomplished musician and a poet. He is best remembered for his ability to depict the beauty of South Georgia's landscapes in poems like *The Marshes of Glynn*.

Intrigued, I looked for other writers and found three more, all born in the early 1900s: Erskine Caldwell, Carson McCullers, and Flannery O'Connor. Caldwell is represented in the Smithsonian by a portrait painted when he was a young man. McCullers, mugging for the camera, and O'Connor, smiling shyly, are captured in black-and-white photographs. Both of them not only shared a South Georgia childhood and talent for writing. They each suffered from health issues that began early in their lives. McCullers had rheumatic fever at fifteen and experienced strokes that left her left side paralyzed. She died at fifty. O'Connor was diagnosed with lupus at twenty-seven and died when she was thirty-nine.

At the time I was occupied with other topics for this book, so I filed away the accidental discovery of four South Georgia writers at the National

Portrait of Erskine Caldwell by Frederick Wight, 1934. National Portrait Gallery collections.

Portrait of Carson McCullers by Irving Penn, 1950. National Portrait Gallery collections.

Portrait of Flannery O'Connor by Joseph Reshower, 1961.
National Portrait Gallery collections.

Portrait Gallery for later use. When I got back to thinking about writers, I decided to give the Smithsonian Libraries a try to see if I could find more South Georgia writers of the same generation as Caldwell, McCullers, and O'Connor. It took a while, but using digital searches I discovered three more who fit my time frame, although their presence in the collections was only in the form of articles in edited volumes.

Two of these, Frank Yerby and John Oliver Killens, were among the first African American writers to gain national recognition for their novels. Raised in South Georgia homes, they left to establish their writing careers. The third discovery was Savannah native Conrad Aiken, the first

Georgia-born author to win a Pulitzer Prize, which he was awarded in 1930 in recognition of his soulful poetry. Tragedy struck early in his life: he was eleven when his father killed his mother and then shot himself. Aiken was taken to live with relatives in New England, where he found solace in poetry. After graduating from Harvard, he wrote and taught at several universities while living alternately in England and the United States. He returned to live in his hometown of Savannah for the last years of his life.

As South Georgia authors, Caldwell, McCullers, O'Connor, Yerby, and Killens all drew on their personal experiences with Jim Crow laws as they took on issues deriving from the poverty and racism of their time. The South then was a place where white Americans constructed a society and legal system based on separate water fountains, seats on the bus, schools, churches, doctors' office entrances, and jails for black and white Americans. Jim Crow could not stand for long, and South Georgia writers helped to expose its hypocrisy through novels that captured the conflicted lives of both blacks and whites caught in a system that pulled everyone down. While segregation affected all Georgians, poverty was not far behind in its impact. In the early decades of the twentieth century, more than 30 percent of farms in Georgia were operated by sharecroppers. Their plight was documented in the book *You Have Seen Their Faces* (1937) by Caldwell with photographs by Margaret Bourke-White. Caldwell and Bourke-White married and are captured in a photograph in the collections of the National Portrait Gallery; their book is in the Smithsonian Libraries.

One South Georgia writer of note whom I could find nowhere in the Smithsonian collections is Caroline Miller. In 1934 she became the first Georgian to win a Pulitzer Prize for fiction with her novel *Lamb in His Bosom* (1933), which relates the lives of poor white farmers seeking to establish homes in South Georgia in the 1820s. Given the book's vivid descriptions of the natural environment of South Georgia and its acclaim, it seemed a perfect fit for the collections. I decided to correct this omission by donating an early edition of *Lamb in His Bosom* to the Smithsonian Libraries.

In all, my search of the Smithsonian collections led to two poets and five fiction writers who spent their childhoods in South Georgia and whose works influenced the literature of the region. Growing up in South Georgia shaped the stories they told. Their writings help us understand the lives of people, black and white, who struggled to make a life in a place where change came slowly and possibilities were limited by poverty and segregation.

## Artists and Art

I came to this part of my search brimming with confidence. I had already seen works in Smithsonian art museums by artists from Georgia whom I admired, including Mattie Lou O'Kelley, Howard Finster, and Jasper Johns. I hoped to include all of them in my book, but this balloon was soon punctured. I found that O'Kelley and Finster lived in North Georgia, not South Georgia, and that Johns, while born in Augusta, lived there less than a year before his family moved to South Carolina.

Chastened, I had to rethink my approach. Friends at the Smithsonian American Art Museum—Deputy Director Rachael Allen and curators Leslie Umberger and Virginia Mecklenberg—came to the rescue. I also spent time reviewing the collections of the National Postal Museum, which contain artwork that was the basis for stamps; the Archives of American Art; and the Department of Anthropology of the National Museum of Natural History. In the end, the art I found provided another view into the place I call home.

### *Landscapes*

South Georgia does not offer striking vistas of snow-covered mountains or sweeping plains, but the land does possess a beauty of its own. George Inness, a prominent landscape painter in the United States in the nineteenth century, was one of the first to recognize it. Inness began visiting Florida and South Georgia in the late 1800s to escape the winters of his northern home. By this point in his long career, his work had taken on an impressionistic style. His painting *Georgia Pines*, completed in 1890 and housed in the Smithsonian American Art Museum, captures a picnic in a longleaf pine barren, its floor carpeted by pine needles and the rays of afternoon sun brightening the dark background of the forest. If you have ever been in a pine barren on a sunlit day, you know there is a stillness and grandeur about it. *Georgia Pines* brings the viewer into this moment.

South Georgia landscapes have attracted painters other than Inness, but their works are not in the Smithsonian collections. However, the Archives of American Art contain the papers of Hattie Saussy and Nell Choate Jones. Saussy, a Savannah native, painted strikingly delicate scenes of the marshes and wetlands near her home city. Jones was born in Hawkinsville, Georgia, but lived most of her life in New York. She began painting later in life, and her work harked back to her childhood home. Her painting *Georgia Red Clay* uses vivid colors and a winding country road to give life to what would otherwise be ordinary farm buildings.

George Inness, *Georgia Pines* (1890). Smithsonian American Art Museum collections.

### *Folk Art*

While North Georgia has Mattie Lou O'Kelley and Howard Finster, South Georgia has Ulysses Davis. Born in Fitzgerald in 1913 to a railroad family, Davis dropped out of school after the fourth grade to become a blacksmith for the railroad. Eventually the company moved him to Savannah, but he was laid off, and, shifting gears, he established himself as a barber. Along the way his boyhood talent for whittling and carving grew into wood sculpting, and the shelves of his barbershop served as his gallery.

The Smithsonian American Art Museum collections contain nine works by Davis, illustrating the breadth of his intellect and imagination. From the ominous *Beast Going through the Grass* to an untitled peaceful South Georgia farmstead, his work covers a wide range of topics. He was influenced by the space program and by his African roots, but in all instances his own character is evident. A South Georgia original, Davis was recognized toward the end of his life for his art, but according to an article in the *New York Times*, he rarely sold any pieces, saying, "They are part of me. If I sold them I would be poor."

Ulysses Davis, *Beast Going through the Grass* (1984–1985). Smithsonian American Art Museum collections.

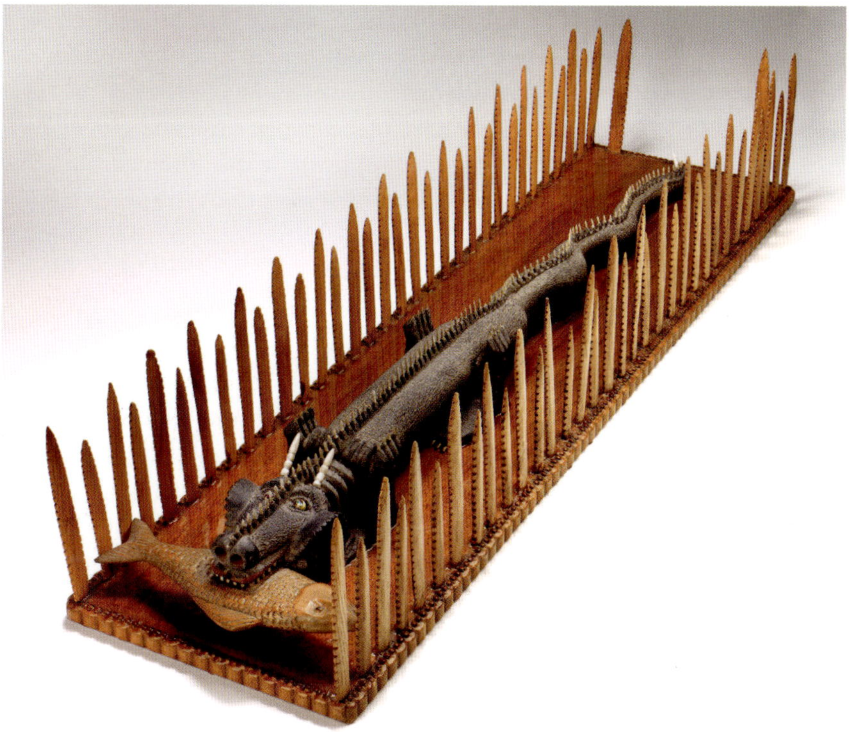

### *Contemporary Art*

Determining who fit my parameters and who did not was not always clear-cut, especially for contemporary art. Two well-known artists who came close were Lamar Dodd and Walker Evans, but neither was born in South Georgia. Both have a few works about my home in the Smithsonian collections, but these are not among their best. Two artists who did meet the test are Alma Thomas and Lamar Baker, both of whom lived through the Jim Crow era in South Georgia.

With thirty paintings in the Smithsonian American Art Museum collections, Alma Thomas is well represented. She was born in 1891 into a middle-class African American family in Columbus, Georgia. In 1907 her parents moved to Washington, D.C., to escape escalating racial violence and find better schools for their daughter. Thomas earned her bachelor's degree in fine arts from Howard University and a master's degree from Columbia University. She made a living as a teacher and worked on her art in her spare time. Upon retiring from teaching in 1960, she made a life-changing decision to paint in an abstract style with strong colors and patterns. As stated in Thomas's 1978 artist biography at the Smithsonian American Art Museum, her childhood in Columbus gave her an abiding appreciation for the vibrant natural world: "The family lived in a Victorian house high on a hill overlooking the town where Thomas spent her childhood observing the beauty and colors of nature." Much later in life she was moved by the early days of the space program. A number of her works, such as *Atmospheric Effects I*, *Atmospheric Effects II*, and *The Eclipse*, show this influence.

Alma Thomas's first large-scale exhibition was at the Whitney Museum of American Art in New York when she was eighty-one years old. After her death her paintings were shown in the public spaces of the White House during the Obama administration, the first time an African American woman was so honored. Thomas was proud of her connection to the Smithsonian, and in addition to the paintings she donated, her papers, photographs, and sketchbooks are part of the collections of the Smithsonian Archives of American Art. Throughout her life she never lost interest in teaching, and she held workshops at the Smithsonian for the children of Washington, D.C.

Seen in person, her works make a strong impact. The colors and patterns of paintings like *Fall Begins* and *Spring Grass* are alluring and offer a sense of the joy of nature. The collections also contain a portrait of Thomas painted by Laura Wheeler Waring in 1945, which shows a strong, dignified

Alma Thomas, *The Eclipse* (1970). Smithsonian American Art Museum collections.

Portrait of Alma Thomas by Laura Wheeler Waring, 1945. Smithsonian American Art Museum collections.

woman who seems to be ready to break into a smile. That is the way I like to think of her.

Lamar Baker was born in Atlanta in 1908, worked in New York City, and then spent the final years of his life in Waverly Hall, Georgia, a small town in the suburbs of Columbus. Baker studied at Atlanta's High Museum of Art while in high school and moved to New York City to join the program of the Art Students League. In 1942 he won a fellowship to travel throughout the South, where he painted scenes related to African American spirituals. Over time his work, which he took care not to advertise, took on a tone of social criticism, with paintings depicting the inhumanities of slavery (*The Slave Plant*) and the debilitating impact of work in a textile mill (*Textile Tangle*). Baker's art can be found in the collections of the Columbus Museum, and there are twenty-three works at the Smithsonian American Art Museum.

*Portraits*

Many people think of a portrait as a painting, but that is not necessarily so. Formally speaking, a portrait is a two-dimensional image of a person, which can be a photograph, drawing, engraving, or painting. But Smithsonian collections, such as those at the National Portrait Gallery, also include sculpture. Portraits in the largest sense of the word are found in a number of Smithsonian museums.

I have already cited the National Portrait Gallery's portrayals of such South Georgians as Revolutionary War hero General Nathanael Greene; Creek leaders William McIntosh, Osceola, and Menawa; writers Joel Chandler Harris, Sidney Lanier, Erskine Caldwell, Carson McCullers, and Flannery O'Connor; and President Jimmy Carter. But the list does not end there. The gallery holds portraits of five men with South Georgia roots who played a role in the Civil War. Four sided with the Confederacy: William Hardee, Lucius Quintus Cincinnatus Lamar II, Herschel Johnson, and Howell Cobb. Hardee was a lifelong military man who served first in the U.S. Army and then as a general in the Confederate Army. Lamar, Johnson, and Cobb were active in national politics before the Civil War, and both Johnson and Cobb served as governor of Georgia. Johnson gained national notoriety in the 1861 presidential race as Stephen Douglas's running mate. He was added to the ticket to provide balance to Douglas's northern base. Of the four Georgians who supported the Confederacy, Lamar most successfully reintegrated into national politics; he became U.S. secretary of the interior and then served as a justice of the U.S. Supreme Court. The other Georgian whose portrait is found in the National Portrait Gallery, John C. Frémont, gained fame as an explorer, politician, and gold miner. He lived most of his life in California, becoming one of the first senators from that state in 1850. In 1856 he ran unsuccessfully for president. He chose to fight with the Union in the Civil War and was named commander of the Western Armies. There are no fewer than thirty-one portraits of Frémont in the National Portrait Gallery collections.

Notable people born in South Georgia who made their mark on the twentieth century and have portraits in the National Portrait Gallery include Juliette Gordon Low, founder of the Girl Scouts; Jim Brown, a football player; Ossie Davis, an actor; Jody Powell, a White House press secretary; performers James Brown and Ray Charles; and Supreme Court justice Clarence Thomas.

Outside of the National Portrait Gallery, the largest collection of portraits at the Smithsonian is in the National Museum of Natural History. The most prolific producer of these works was De Lancey Gill, who worked for the Bureau of American Ethnology from 1890 to 1930. He took more than two thousand photographs of Native Americans, including Geronimo and Chief Joseph. Along the way he photographed dozens of young Muscogee men and women who made their homes in Georgia, Florida, and Alabama. It is not possible to determine which of the people in Gill's photographs lived where, but it is likely that more than a few were from South Georgia. While these photographs were taken for scientific purposes, Gill showed his subjects in dignified poses. Their faces are those of a proud people who have known defeat.

### *Public Art*

During the Great Depression, artists were paid by the federal government to provide art for communities and public buildings. When it came to art for post offices, the Treasury Department was in charge. Through its Section of Painting and Sculpture (also known as the Section of Fine Arts), it saw to the creation of more than eleven hundred murals and three hundred sculptures. Thirty-seven Georgia post offices were chosen for art projects, nineteen of them in South Georgia. Douglas for some reason did not make the cut.

Most of the artists who participated in the post office art projects in South Georgia were from the North, and they were encouraged to make an effort to get to know the towns where their art would reside. Studies by two of these artists are in the collections of the Smithsonian American Art Museum.

Daniel Putnam Brinley was fifty-nine years old when he was commissioned to paint a mural for the post office in Blakely, Georgia, a town in the southwestern corner of the state. There, English settlers contended with the Creeks in the 1700s and 1800s for control of the land. In some cases the settlers won concessions through battle and in others through purchase. Brinley's mural is titled *The Land Is Bought from the Indians*, and stylistically it seems to place the purchase between the settlers and the indigenous people in the mid- to late 1700s. The work is a departure from Brinley's normal style, which ranges from impressionistic to modern, but its bow to history seems to fit its post office setting perfectly. The mural, produced in 1938, still hangs in the Blakely post office.

Any mention of the South Georgia town of Vidalia brings to mind the famous sweet onions that grow in the sandy soils of the area. It was there that thirty-six-year-old Daniel Celentano, born in New York City, painted

Daniel Celentano, *The Country Store and Post Office*, study for mural in Vidalia, Georgia, post office, 1938. Smithsonian American Art Museum collections.

a mural for the post office. A pupil of Thomas Hart Benton, he adopted his usual approach of depicting everyday scenes with a strong sense of realism. The mural, titled *The Country Store and Post Office*, shows an animated gathering of men catching up on the news in the center of a small southern town in the 1800s. Off to the side of the men and their merriment, a solemn woman stands with her children as she reads a letter delivered to her by the horse-drawn stage that is leaving town. Seen today, the painting evokes nostalgia for a time when a small town was a place where people of all walks gathered and created a community.

Unfortunately, the Celentano mural was damaged during renovations to the post office, but a part remains on view. The entirety of the image can be appreciated in Celentano's preliminary study for it, which is in the collections of the Smithsonian American Art Museum. In almost all other cases the works created during the Great Depression remain on display in small-town post offices in South Georgia, reminders of the importance of art in our lives and our history.

## Music and Musicians

New forms of American music exploded on the southern scene in the early 1900s. Musicologists generally explain that jazz came from New Orleans, rhythm-and-blues from the Mississippi Delta, and country music from the Appalachian Mountains. That allocation pretty much leaves out South Georgia, an omission this book will set right.

My search for music that originated in South Georgia took me to the Smithsonian Center for Folklife and Cultural Heritage and its Folkways World Music Collection. According to its website, Folkways Music owes its existence to the passion of Moses Asch, whose goal was to "document people's music, spoken word, instruction and sounds from around the world." The recordings do not focus on those who were already famous, but rather on artists who were creating something new and different. Along the way Asch made or collected early recordings of artists such as Woody Guthrie, Pete Seeger, and Lead Belly. When Asch died in 1986, the Smithsonian acquired the Folkways Music recordings and has added extensively to the collection since that time.

At Folkways I found help from archives director Stephanie Smith and archivist Jeff Place, whose encyclopedic knowledge gave me information or leads that allowed me to chase down South Georgia musicians in the music collections. I did have to help them understand what I was looking for, since recordings are not stored in categories like "South Georgia." But once they got the idea, they came up with an impressive list of possibilities.

### *Jazz and Blues*

At the turn of the twentieth century, South Georgia was a place with hot weather, a depressed economy, a hangover from the Civil War, and an abiding tradition of music rooted in the Gullah Geechee culture—all the right ingredients needed to foster music from blues to country. Place and Smith suggested that a starting point to understand how South Georgia nurtured new music forms could be recordings by Ma Rainey, many of which are in the Folkways archives. I learned that Ma Rainey was the stage name for Gertrude Malissa Nix Pridgett, born and raised in Columbus, Georgia. As early as 1900 she was beginning to develop a style of singing slow-beat blues vocals like nobody else. She was so good she eventually became known as

Gertrude "Ma" Rainey, *Queen of the Blues*, Biograph Records, 1971. Smithsonian Folkways collection.

*Blind Willie McTell, 1927–1935*, Yazoo Records, 1991. Smithsonian Folkways collection.

"Mother of the Blues." Many of her songs, such as "Black Bottom" and "Bo-Weavil Blues," are considered classics.

Ma Rainey began performing in 1904, and her reputation grew; she traveled to New Orleans in 1914, meeting Louis Armstrong and other jazz greats who were creating a new art form. She became a fixture, traveling with Armstrong, and in 1923 she made her first recording, extending her recognition beyond the South. Her records are not only in the Folkways collections but also in the National Museum of American History, which has a 1924 Paramount record of "Lost Wandering Blues." Listening to her early recordings was a revelation. She was that good.

Ma Rainey retired to Columbus in the mid-1930s and ran local theaters until her death in 1939. In recognition of her contributions to our national culture, a likeness of Ma Rainey appeared on a 1994 stamp issued by the U.S. Postal Service, which is part of the collections of the Smithsonian's National Postal Museum. She was a true American original, a mentor to other young singers, and a jazz and blues singer before anybody knew what jazz and blues were supposed to be.

Two musicians from South Georgia are known for their work in the Piedmont blues style, in which fingerpicking guitar players produce musical rhythms akin to ragtime piano. Piedmont blues originated in the Piedmont plateau area of Virginia, the Carolinas, and Georgia. By the 1920s musicians

Blind Willie McTell and Bumble Bee Slim (born Amos Easton) had made their mark. McTell, using a twelve-string guitar, created a sound that was distinctive when compared to the traditional Piedmont blues style. He wrote much of his own music, and two of his songs, "Statesboro Blues" and "B and O Blues No. 2," became national hits. The Allman Brothers Band, Taj Mahal, and Bob Dylan are among the later artists who cited the influence of Blind Willie McTell on their work and recorded versions of his tunes. Recordings of McTell and Bumble Bee Slim are part of the Smithsonian Folkways collection.

### *Country and Popular Music*

Beyond jazz and blues, the city of Columbus shows up again on the country side of the ledger. Tom Darby and Jimmy Tarlton, performing as Darby and Tarlton, wrote a string of hits in the 1920s that have become part of the country and folk lexicon today. Their songs "Birmingham Jail" and "Columbus Stockade Blues" have been recorded by just about every country singer worth his salt. The lyrics of "Columbus Stockade Blues" are a poor boy's lament over being thrown into the clink:

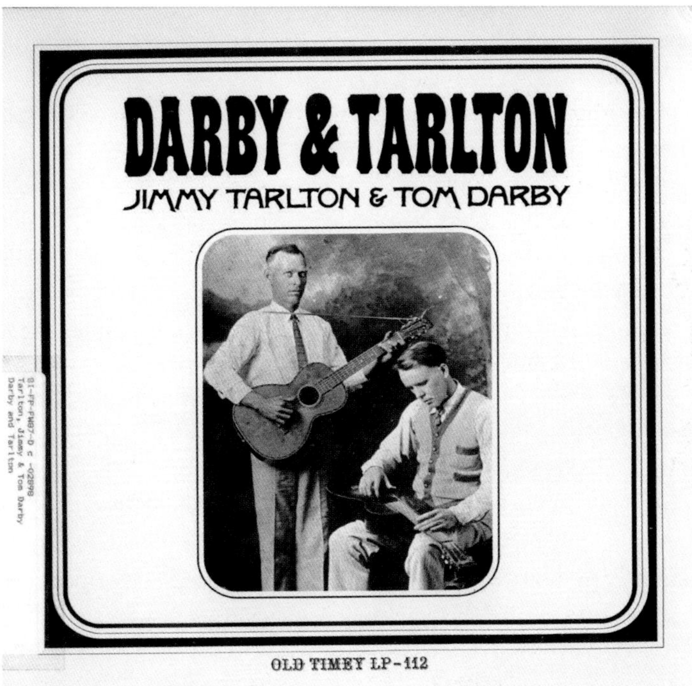


*Darby and Tarlton*, Old Timey Records, 1971. Smithsonian Folkways collection.

Sheet music for "Lazybones," music by Hoagy Carmichael and lyrics by Johnny Mercer, 1932. National Museum of American History.

Way down in Columbus, Georgia,
Lord I wish I was back in Tennessee.
Way down in that old Columbus Stockade,
My friends all have turned their backs on me.

Smithsonian Folkways has recordings of "Columbus Stockade Blues" by different artists in its collections. I personally recommend the version sung by Doc Watson.

Popular music cannot be left out because of Johnny Mercer, who was born in Savannah in 1909. He brought his lyrics and songs to Broadway and Hollywood, winning four Oscars along the way. As a lyricist he had legendary collaborations, writing his first big hit, "Lazybones," with Hoagy Carmichael. The two of them also worked together on "In the Cool, Cool, Cool of the Evening," which won an Oscar in 1951. As a lyricist working with Henry Mancini, Mercer won Oscars for "Moon River" and "Days of Wine and Roses," and with Harry Warren he won an Oscar for "On the Atchison, Topeka, and the Santa Fe." Mercer was influenced by his upbringing in Savannah, where he attended black churches and listened to the fishermen at the wharf who sang in Gullah.

I found recordings of Mercer and his songs in Folkways Music, but also in other Smithsonian collections. Some of his records, conductor's scores, copy scores, and the sheet music for "Lazybones" are in the collections of the National Museum of American History. In addition, the National Postal Museum collections contain a stamp issued in 1996 with his image.

### *Soul*

No discussion of musicians from South Georgia in the Smithsonian collections can be concluded without mentioning two artists whose work transcends categories: Ray Charles and James Brown. Both faced challenges in their youth and overcame them with talent and determination.

Ray Charles Robinson was born in 1930 in Albany, Georgia, where his mother, Aretha Williams, had family. However, his mother actually lived with the Bailey Robinson family in Greenville, Florida, where she worked as a sharecropper. Bailey Robinson, Charles's father, eventually left the family, and Charles was raised by Aretha and the matriarch of his father's family, Mary Jane Robinson. It is reasonable to ask how Ray Charles qualifies for this book since he grew up in Florida. But he makes the cut not only because he was born in South Georgia, but also because his 1960 recording of the song "Georgia on My Mind" by Hoagy Carmichael and

Ray Charles at the piano, 1961. Photograph by Michel Salou. National Portrait Gallery collections.

Stuart Gorrell became the state song of Georgia in 1979. My personal appreciation for him traces back to his performance to a sold-out crowd in Alexander Memorial Coliseum at Georgia Tech in 1963 when I was a student there. All of us in the audience could not get enough of Ray Charles that evening.

Charles, blind by age seven, was educated at the Florida School for the Deaf and Blind, where his natural musical talents were nurtured. His interest in jazz led him toward popular music, and he finally got his big break in Chicago. A master songwriter, singer, and performer, his career took off with the 1960 album *The Genius Hits the Road*, which includes the iconic track "Georgia on My Mind." Charles and his career are represented in the Smithsonian collections in the National Museum of American History, the National Museum of African American History and Culture, the National Portrait Gallery, and the Smithsonian Libraries and at the Cooper Hewitt Design Museum in New York City.

It may be hard to top the challenges Ray Charles overcame to find his place in music, but James Brown had his own very steep hill to climb. He was born to a poor family in South Carolina in 1933, then moved to Augusta, Georgia, with his mother. They lived in a roadhouse where young James developed his talent for music. At sixteen he was sentenced to eight to sixteen years in prison for robbery, but he formed a gospel music group and won parole after less than four years.

With help from Little Richard (Richard Wayne Penniman), who is from Macon, Georgia, and was by then a rising rock-and-roll star, Brown came into his own. He was a singer and performer, a songwriter and producer. Known for his commitment to performance, he earned the title of "Mr. Dynamite" and later "Godfather of Soul." His songs "I Feel Good," "Please, Please, Please," and "Living in America" are ingrained in American life.

James Brown's talents landed him in the Smithsonian collections in portraits and photographs in the National Portrait Gallery and artifacts housed in the National Museum of African American History and Culture. The latter collections include jumpsuits (red and black) he wore to perform, a black cape, and the lyrics for the song "Everybody's Got a Thing," written in his own hand on yellow legal paper.

Little Richard and his fellow Macon resident Otis Redding are two other South Georgia musicians who were pioneering soul, rhythm-and-blues, rock-and-roll songwriters, singers, and performers. Redding's career was cut short when his plane crashed in 1967, but he had already blazed a trail for a generation with ballads that still resonate. His first big hit, "These Arms of Mine," was released in 1962 and began his relationship with Stax Records and Steve Cropper, a musician and record producer. His last hit, "(Sittin' on) The Dock of the Bay," was

Red jumpsuit owned by James Brown, 1970–1989. National Museum of African American History and Culture collections.

released after his death and was cowritten with Cropper. In between these two songs he established a body of work that continues to be recorded by artists today. Many of his records are housed in the Smithsonian Folkways collections, and the original contract for the hit "Mr. Pitiful," signed by Cropper and Redding, is in the collections of the National Museum of African American History and Culture. Redding was commemorated in 1993 by a U.S. postage stamp, which is in the National Postal Museum, and his biography by Mark Ribowsky, *Dreams to Remember* (2016), is in the Smithsonian Libraries.

Otis Redding gained fame when I was in college, but Little Richard burst on the scene with "Tutti Frutti" in 1955 when I was a freshman in high school. More than a hit, it just blew us away. Nobody had ever heard a song with the lyric "A-wop-bop-a-loo-bop-a-wop-bam-boom!" In 2010 when "Tutti Frutti" was added to the U.S. Library of Congress National

*Little Richard's Grooviest 17 Original Hits!*, Teichiku Records, 1968. Smithsonian Folkways collections.

Recording Registry, the library's announcement said that the song had "announced a new era in music." Many of Little Richard's recordings, including his 1968 album, *Little Richard's Grooviest 17 Original Hits!*, are among the collections of Smithsonian Folkways. A biography, *The Life and Times of Little Richard: The Quasar of Rock* (1994) by Charles White, is in the Smithsonian Libraries.

The rock-and-roll, jazz, rhythm-and-blues, soul, and country musicians who lived in and breathed the air of South Georgia made their marks not only through performance but also through the songs and lyrics they wrote. Their creative legacy contributed to the music we know today as genuinely American. Although I am not a musicologist, I believe that the role South Georgia musicians played in creating our musical heritage has been underestimated by historians.

## The Moving Power of the Arts

Going in, I knew that the art collections of the Smithsonian were not as comprehensive as those in the sciences, but I felt that just about anything I could find might offer insights into the cultural history of South Georgia. The collections did not disappoint. I learned that South Georgia was home to writers Erskine Caldwell, Carson McCullers, Flannery O'Connor, Frank Yerby, John Oliver Killens, and Conrad Aiken; painters and sculptors Alma Thomas, Lamar Baker, Ulysses Davis, and Hattie Saussy; and musicians Ma Rainey, Blind Willie McTell, Ray Charles, James Brown, Little Richard, Otis Redding, and Johnny Mercer.

These remarkable people with South Georgia roots made their mark on our nation's culture and our consciousness. They still speak to us through their words, art, and music: painters who show us the beauty of South Georgia from its sea-tossed coast to the quiet of a pine barren; writers who strip away the civilized veneer of a society to show us the lives of those damaged by racism and poverty; and musicians who fill our spirits with song and our lives with joy, even when the lyrics are at times stories of tragedy.

Collectively these artists also speak of the courage of those who would not be broken, no matter the odds. These voices have illuminated the way for future generations.

CHAPTER FOURTEEN

# Jimmy Carter

James Earl Carter Jr. is the only person born in South Georgia, or all of Georgia for that matter, to have served as president of the United States and won the Nobel Peace Prize. Either of these remarkable achievements alone would have secured his place in these pages because they guarantee that he is represented in the Smithsonian collections. But neither is the principal reason his story is the last one in this book. It has to do with his bond to the land that is South Georgia and its people, a bond he and I share. It is his life, which spans the passage from an era dominated by poverty and racism to an era in which educational opportunity exists and legal segregation has been relegated to the past. A life that connects to my own.

Jimmy Carter is typically viewed through the lens of his presidency or his post-presidential role in promoting peace and eradicating disease. Dozens of books and articles have been written about these subjects by prize-winning authors, including Carter himself. My purpose is more modest: to trace his work and accomplishments back to his home in South Georgia, where his love of the land and nature shaped him. As he says in his book *An Outdoor Journal*, "I have never been happier, more exhilarated, at peace, rested, inspired, and aware of the grandeur of the universe and the greatness of God than when I find myself in a natural setting not much changed from the way He made it." Anyone who ever wondered why Jimmy Carter, as president, moved to protect more than a hundred million acres of land in Alaska need look no further than South Georgia, where he grew up working on a farm and fishing and hunting in nearby streams and forests.

Artifacts connected to Carter are in six Smithsonian collections, and the objects include books, medals, photographs, paintings, sculptures, political cartoons, campaign memorabilia, an inaugural gown, peanuts, a solar panel, and a blog about birds. Despite the number and variety of objects on this impressive list, two things should be noted. First, the Jimmy Carter Presidential Library and Museum in Atlanta holds a far more comprehensive collection about him and his family than the Smithsonian does. Second, if you were to plot the Smithsonian's objects against time, the result would be a bell curve, with the maximum number centered on the period when he served as president. However, as is often the case, it is the outliers that are most interesting.

## The Early Years

To understand Carter, I begin with his childhood. While the Smithsonian collections have no artifacts from that time, they include books about it. The collections of the Smithsonian Libraries and the National Museum of American History include six biographies of Jimmy Carter written between 1979 and 1997. The same collections include seven books written by Jimmy Carter and one by Rosalynn Carter titled *First Lady from Plains*. Jimmy Carter's books range from *A Government as Good as Its People* (1977) to *A Full Life: Reflections at Ninety* (2015). However, Carter has written thirty-three books, so the seven in the Smithsonian are only a sampling of the total.

Most of the books in the collections related to Carter, or written by him, focus on weighty topics, such as governing the nation, negotiating peace agreements, dealing with world leaders, and eradicating disease. The one book that does not fit the mold is *An Hour before Daylight: Memories of a Rural Boyhood* (2001). I do not know why it was chosen to be in the Smithsonian, but it is fortunate for me that it was. Reading it, I was struck by how similar his childhood was to those of my parents and myself and how this grounding was crucial to what he did later in life.

My mother and father were born into South Georgia farming families in 1902 and were twenty-two when Jimmy Carter was born in 1924. In those days, times were hard for both blacks and whites living on farms, with no electricity or running water in homes and few educational opportunities. To endure, especially through the Great Depression, families relied on love, commitment, and hope. Although my parents had to leave South Georgia in the 1920s, they found a way to return and create a new life. As a result, the next generation of my family, including me, experienced childhoods in South Georgia.

Carter, like myself, was lucky to have loving, hard-working, dedicated parents. Lillian and James Earl Carter Sr. provided for their children as best as a farming family in rural Georgia could while exemplifying the fundamental values that would guide their lives. Jimmy Carter's family gave him a long leash, allowing him to spend time exploring nature and encouraging his independence and willingness to stand up for his beliefs.

Carter's descriptions of his outdoor adventures as a boy could almost be mine. He enjoyed nature, hunting doves and quail, and fishing. I was not surprised when I read in *An Hour before Daylight* that his favorite fishing experiences happened when he and his father drove to the Upper Satilla River and the Okefenokee Swamp. As he put it, "The Little Satilla River in Southeast Georgia was my favorite place for fishing. We'd wade waist deep out into the river, on sandbars, and fish along the undercut opposite banks on the outside of curves in the meandering stream. It was a remote and lonely site, so Daddy and I stayed fairly close together. We fished with pond worms and caught mostly large bream—redbreast, warmouth, and bluegills." The only difference between his experiences and mine was that he had to travel farther to get to the Satilla River than I did.

Carter's love of the land came from his direct connection to it: "My most persistent impression as a farm boy was of the earth. There was a closeness, almost an immersion, in the sand, loam, and red clay that seemed natural, and constant. The soil caressed my bare feet and the dust was always boiling up from the dirt road that passed fifty feet from our front door, so that inside our clapboard house the red clay particles, ranging in size from face powder to grits, were ever present." Feeling the presence of the soil and seeing it tilled and nurtured to provide the food of daily life built a sense for the land that would evince itself later in life in Carter's commitment to the environment and wilderness.

His understanding that the land had a history of its own grew as he and his father became committed collectors of Native American artifacts they found on their property: "From the time I was five years old, my father and I were fascinated by Indian artifacts and the lives of the Native Americans who had inhabited the western part of Georgia. Knowing that our family's farmland had once been theirs and that our ancestors had confronted and replaced them in this area deepened our interest in their history."

In the course of writing this book, I visited the Carter Museum, where the curatorial staff was kind enough to show me a half dozen trays containing hundreds of projectile points and pottery sherds that Carter and his family collected over time. Given Carter's organized mind, it was not surprising to find the collections carefully documented with the initials of the person who found each piece and a system to identify the location.

Experts had recently reviewed the collection and classified each object as to its approximate date and the name of its archaeological period. These names—Mississippian, Woodland, Archaic—are the same as those cited in chapter 10 of this book. Some of the points date back to 10,000 BCE. I was struck by several of the oldest pieces, unusual because they were created in quartz, a material that resists carving because of its hardness and lack of cleavage planes. Gleaming white, beautiful, and exhibiting the crystalline structure of quartz, they are the work of masters who lived ten thousand to twelve thousand years ago. Creating this collection had a formative impact on Carter that would be reflected later in his actions as president.

Carter's educational journey took him away from Plains for a time. His father, known as Earl, supported his son's desire to go to the U.S. Naval Academy in Annapolis and worked every angle he could think of to get Jimmy admitted, including lobbying politicians. In spite of all his work, the appointment had not come when Jimmy graduated from high school in 1941. Instead, after a stint at the local junior college, he was admitted to study engineering at my alma mater, the Georgia Institute of Technology. Patience proved to be the key as Carter was appointed to the Naval Academy the next year, although he had to start over as a freshman. He graduated and ultimately became the senior officer on one of the pre-commissioning crews of the second nuclear submarine in the U.S. Navy fleet.

## The Budding Politician

When Earl Carter died, Jimmy Carter resigned his commission in the navy and returned to Plains to run the family farm and business. Even as he was learning firsthand how hard it is to make a living farming, he found time to get involved in local and then state politics, and these experiences relate to the first object to arrive in the Smithsonian collections about Jimmy Carter.

In 1978, one year after Carter became president, *Time* magazine decided to give the Smithsonian's National Portrait Gallery its collection of the artwork that had been created for its covers. Among the pieces was the cover art for the May 31, 1971, issue, showing a young Jimmy Carter who had just been elected governor of Georgia. He is smiling, and in the background behind him are overlapping flags, one of the Confederate States of America and the other of the United States. The headline reads, "Dixie Whistles a Different Tune."

The article related to the cover is about a group of southern governors who had taken office in 1971 on progressive platforms. They all shared a belief that it was time to move on from the days of segregation and concentrate on improving the economy and providing educational opportunities to all citizens. While other governors were included in the article, Jimmy Carter was on the cover and his was the most compelling story.

## Running for President

At the time, the Georgia Constitution limited the governor to one four-year term. By the time Carter completed his term in January 1975, plans for his campaign for the presidency were under way. His lack of national name recognition was a challenge, and he quickly had to make a case for himself as a fresh face with substance so he could win enough early primaries to stay in the race. The Smithsonian collections demonstrate how he managed it.

### *Selling His South Georgia Roots*

When a political campaign is finally over, the last job of the campaign staff is to clean out the headquarters, usually by throwing away what remains. This moment is when the Political History Division of the Smithsonian's National Museum of American History goes to work. For more than fifty years this division has collected campaign memorabilia, in part by swooping into closing campaign headquarters and scooping up buttons, posters, and pamphlets before they can be chucked out by the staff.

As Harry Rubenstein, curator for the Political History Division, began pulling out cabinet drawers labeled "1976 campaign," I learned that South Georgia peanuts, not the styrofoam kind used today for package shipping, were a large part of the Carter campaign strategy. Carter ran as an outsider, and one way to reinforce it was to emphasize his experience as a real peanut farmer. His campaign props included a simple, large, heavy-duty bag used by Carter's Warehouse of Plains, Georgia, to ship fifty pounds of seed peanuts. The bag's label said the peanuts were "hand picked and custom shelled." The message was that this man Carter was the real deal, not a gentleman farmer.

There also were posters featuring large images of peanuts, often drawn with a wide toothy smile to reinforce the connection to the candidate. The peanut imagery was extended into three dimensions through two-foot-high peanut banks made of yellow plastic with, yes, a wide toothy smile.

Beyond the message that he was a man of the soil, Carter and his team knew they had to introduce him to voters who had not known him as

Carter for President campaign peanut bank. National Museum of American History collections.

governor of Georgia. Thus was born the "Hello, I am Jimmy Carter and I am running for president" theme, which cropped up on buttons and posters everywhere. The Smithsonian has a number of examples, and as an early Carter supporter, I had one of the buttons myself. I also had a "Carter for President" sticker on the rear window of my station wagon. None of it helped, because I was living in California at the time, and Carter lost the California primary to Governor Jerry Brown, a late entry into the race who ran as an "ABC" (Anybody but Carter) candidate.

Carter's strong record on the environment as governor of Georgia attracted environmentalists to his campaign. One of the campaign pins in the Smithsonian collections proclaims "Conservationists for Jimmy Carter,"

Environmentalists for Carter presidential campaign pins. National Museum of American History collections.

while another shows trees, a river, deer, and birds with the message "Why Not the Best? / Carter Mondale '76." A whimsical pin from California declares "Redwoods for Carter," an allusion to a quote attributed to Ronald Reagan (who was competing with Gerald Ford for the Republican nomination) that "if you have seen one redwood tree, you have seen them all." As president, Carter returned the favor by promoting the passage of legislation to protect the redwoods.

The last item Rubenstein showed me from the Carter campaign caused me to laugh, which I am sure was the intent. It was an image of Carter dressed in a spotless white robe and with a long beard and bore the caption "J. C. Can Save America!"

### *The Art of the Campaign*

After Carter won several of the early Democratic primaries, he attracted national attention, and *Time* magazine again took notice. The March 9, 1976, cover featured him riding the Democratic Party mule with his competitors, Mo Udall, George Wallace, and Henry "Scoop" Jackson, tugging backward on the mule's tail. While it did show him in a suit, he was wearing a farmer's straw hat and holding a paper bag of peanuts. The headline was "Who Is Jimmy Carter and Why Do They Love/Hate Him?" By May, Carter had won sixteen primaries, and the May 10, 1976, *Time* cover featured him wearing a plain denim shirt with the headline "Jimmy's Breakthrough."

Having captured enough primary votes to win the Democratic nomination, Carter next appeared on a *Time* cover on June 10, 1976, along with Ronald Reagan and Gerald Ford, who were still competing for the Republican nomination. The headline read, "Our Next President (Pick One)." Carter officially won the Democratic nomination that July and the presidential election, beating Ford, on November 2, 1976.

Some of the best images from the campaign were taken by the photographers who shadowed Carter, including many who were considered the foremost of their time. *New York Times* photographer George Tames was known for taking a different approach to capturing his subjects. In 1975 he

Jimmy Carter by Alan Reingold. This artwork was used for the cover of *Time*, May 10, 1976. National Portrait Gallery collections.

took trips to Plains to photograph Carter at work there, both as a farmer and as a family man. One of the most endearing of these photos is housed in the collections of the National Portrait Gallery. It shows Jimmy Carter holding his daughter, Amy, in the kitchen of their home in Plains while Rosalynn looks on admiringly from the background.

When the primaries were over, the Democratic Party wanted to strengthen Carter's appeal to young voters. Andy Warhol was chosen to create a new kind of image that could be used for fundraising and campaign literature. The end result was a screen print on paper titled *Carter I*, and the *Hope* poster created in 2007 by Shepard Fairey for the campaign of Barack Obama looks strikingly like it. The portrait by Warhol was so successful that Carter declared it helped him win the election; the original is held by the National Portrait Gallery and was signed by both Warhol and Carter.

## The Presidency

Jimmy Carter was inaugurated on January 20, 1977, a cold day in our nation's capital. The most memorable moment occurred when he and First Lady Rosalynn Carter chose to walk in the inaugural parade rather than ride in the armored limousine provided for them. It was a first, and it sent a message to the public that the Carters were cut from a different cloth than those who had gone before them. That inaugural walk was captured in a photograph, which is now in the collections of the National Museum of American History.

It is a common practice to cast a gold medal commemorating a presidential inauguration, and Carter choose a Georgian, Julian Hoke Harris, to design his. Harris was a 1928 Georgia Tech architecture graduate who taught at Georgia Tech for thirty-four years while painting and sculpting on the side. Every morning at breakfast during my freshman year, I looked up at a beautiful stained-glass window he had created for the school's Brittain Dining Hall. The gold medal Harris designed for the Carter inauguration is in the collections of the Smithsonian American Art Museum.

As is the custom, Rosalynn Carter donated her gown and purse from the inaugural ball to the Smithsonian for the First Ladies Collection in the National Museum of American History. The gown is famous because it is one of the first that was not chosen expressly for the occasion. The First Lady had worn the same gown in 1971 for the ball held when Jimmy Carter was inaugurated as governor of Georgia.

### *Magazine Cover Art*

The Carter presidency, like any other, was filled with ups and downs, some of his own making and some caused by events beyond his control. Unfortunately for him, the negative events were weighted toward the end of his term and affected his 1980 reelection campaign. The *Time* magazine cover art collection at the National Portrait Gallery offers a window into the issues that dominated public debate during his presidency. From 1977 to 1980 Carter was featured on fourteen covers, but even when he was not there in person, many magazine covers focused on the issues that faced his administration.

As his inauguration approached, Carter was named Man of the Year by *Time*, and the January 7, 1977, cover showed him dressed in down-home denim against a background of the Plains water tower and his peanut farm. The surprise was that his trademark smile was missing. Carter did not appear smiling on any future *Time* covers, not even on the October 2, 1978, cover that hailed his remarkable achievement in negotiating the Camp David Accords between Anwar Sadat of Egypt and Menachem Begin of Israel.

Events nationally and globally began to work against Carter during the first half of 1979. In Iran the shah, America's long-time ally, was forced to leave the country, and Islamic militants came to power; the Three Mile Island nuclear accident occurred; the Organization of the Petroleum Exporting Countries (OPEC) announced new price hikes, causing long lines at gas stations; and inflation steadily increased. The July 23, 1979, cover of *Time* showed a concerned Jimmy Carter with the headline "At the Crossroads." The magazine included an interview with him about the growing issues the nation faced.

Things did not get better in the second half of 1979. In November, a group of anti-American students overran the U.S. embassy in Iran, taking hostages and holding fifty-two of them for the remainder of the Carter presidency. To top off the year, the Soviet Union invaded Afghanistan. The covers of *Time* magazine reflected these events, showing the blindfolded hostages in Tehran and a Red Army general with his sights on Afghanistan.

In 1980 with the elections approaching, Carter authorized a helicopter raid to free the American hostages in Iran, but it was unsuccessful and eight lives were lost. The *Time* cover for May 5, 1980, featured a somber portrait of the president and the words "Debacle in the Desert."

Carter was in an unenviable position, and his nomination fight with Senator Ted Kennedy weakened him as he approached the general election battle against Ronald Reagan. His last *Time* cover as president came on August 25, 1980, and was headlined "Running Tough." Carter was no quitter and fought hard to win, but he was defeated handily in the election.

### *Official Portraits*

One official portrait of each president hangs in the Hall of Presidents of the National Portrait Gallery. Jimmy Carter's portrait was painted by Robert Templeton, a highly respected artist who had to his credit portraits

Portrait of Rosalynn Carter by Robert Templeton, 1976. National Portrait Gallery collections.

of President Lyndon Johnson and Dr. Martin Luther King Jr. Carter's portrait is dignified, but Templeton's use of pale colors, the lack of the trademark Carter smile, and the somewhat stooped posture stand in contrast to more vibrant presidential portraits in the hall. A later portrait by Herbert E. Abrams, completed in 1982 for the White House collection, more clearly shows his vitality and personal warmth.

Rosalynn Carter's official portrait as First Lady, along with those of other First Ladies, is held by the White House Historical Association, not the Smithsonian. However, the Smithsonian has its own small portrait of her, painted in pastel on illustration board by Robert Templeton. Here, Templeton was at his best, showing the First Lady in two contemplative poses with a hint of a smile. Rosalynn Carter was one of the most active First Ladies, working alongside her husband on the rights of children, refugees, and those with mental illness.

### *Amy Carter and the Elephant*

After I became secretary of the Smithsonian in the summer of 2008, my first trip to see something of the institution beyond the National Mall was to the National Zoo. It was there I met Shanthi, one of two female elephants at the zoo. Shanthi was a gentle, intelligent creature who lived up to the meaning of her name—peace and tranquility. When I met Shanthi in 2008, she was thirty-two years old and had lived at the zoo for all but one year of her life.

During my research on the Carter family and the Smithsonian, a computer search unexpectedly returned Amy Carter's name together with Shanthi's. The connection went all the way back to 1977. When Shanthi was a baby living in Sri Lanka, her mother was killed by poachers, and she was weaned and raised at an elephant orphanage. Coincidentally, the United States was celebrating its bicentennial at the time, and in honor of this occasion Sri Lanka offered to send an elephant to the National Zoo. It took a while to sort out the logistics of shipping an elephant from Sri Lanka to Washington, D.C., but in the spring of 1977 the gift finally arrived—one-year-old Shanthi. Again coincidentally, a new president named Jimmy Carter had just taken office, and his daughter and wife loved animals. A clever zoo director named Theodore Reed invited Amy and Rosalynn to participate in the official welcoming ceremony for Shanthi. So two South Georgians were on hand with dignitaries from the Smithsonian and Sri Lanka to greet the new baby elephant—the same elephant I met thirty-one years later on my first visit to see the

One-year-old elephant Shanthi is greeted at the National Zoo by Amy Carter, Rosalynn Carter, and Punita Gunaratne, April 2, 1977. Smithsonian Institution Archives.

Smithsonian's living collections at the National Zoo. A photograph of the event in the Smithsonian Archives shows a young Amy with her mother and an even younger Shanthi.

## Ahead of the Curve

As the dust surrounding the Carter presidency settled, it took more than the usual amount of time for some of Carter's accomplishments to be appreciated. For example, Carter was ahead of his time in understanding the nation's need for non-carbon-based sources of energy. Having grown up in South Georgia and having been educated as a nuclear engineer, he knew something about the power of the sun.

Housed in the collections of the National Museum of American History is one of thirty-two solar panels that Carter had installed on the White House in 1979 to heat the water used for the laundry and kitchen. The panel, state-of-the-art for the time, is about nine feet long and three feet wide and weighs a hundred pounds. The solar panels did their job at the White House, but the real point was to demonstrate the future of solar energy. At the time Carter said, "A generation from now this solar heater can either be a curiosity, a museum piece, an example of a road not taken, or it can be just a small part of one of the greatest and most exciting adventures ever undertaken by the American people." He was right on all counts. The solar panels served as a visible symbol of Carter's energy policies, which included creating the Department of Energy and the Solar Energy Research Institute in Golden, Colorado, in 1977. He championed tax credits for businesses that created new sources of energy and for homeowners that installed solar panels on their homes.

However, our nation's journey toward the use of alternative fuels has been full of fits and starts, destabilized by inconsistent federal government policies. After Reagan became president in 1981, it did not take him long to dismantle Carter's solar-friendly policies as well as the solar panel array on top of the White House. Even so, the Department of Energy and its laboratory in Golden, Colorado, now named the National Renewable Energy Laboratory, have continued the research needed to bring solar power into the mainstream, and it is now achieving a new level of prominence. Today you can buy solar panels from big box stores like Home Depot and Walmart, and if you look up you can see the roofs of the stores covered by solar panels quietly and efficiently generating power. The evidence for the resurgence is visible in South Georgia, where increasing numbers of large solar arrays are being installed to capture the abundant sunshine that warmed Jimmy Carter and me as boys in the fields. Power generated by these solar units is now cheaper than that provided by coal plants, and tax revenues from the arrays benefit the rural communities where they are installed.

Jimmy Carter was not to be left out of this latest solar revolution in South Georgia. In February 2017, he threw the power switch on a thirteen-acre solar array installed on one of his peanut fields. The power it generates flows into the Georgia Power Company system and helps provide electricity for the city of Plains. Not long after it was installed, I visited Plains and found the field where Carter's solar array stands. No signs lead you there, nor is there a signature steam plume such as a coal plant emits. The panels just sit quietly in a small field behind some residences, doing their job without generating greenhouse gases or toxic waste or using any water. If you are there on a sunny day and stand in place long enough, you will see the panels shift position to track the South Georgia sun as it moves from east to west. Of course, nature developed this strategy millions of years ago, and the leaves of the oak trees surrounding the field are doing exactly the same thing.

## Protecting the Crown Jewels

The National Park Service says that Theodore Roosevelt is known as our "conservation president," and rightfully so, since he protected 250 million acres of land for parks and national forests and promoted the passage of the Antiquities Act, which led to the creation of our national monuments. His interest in conservation was sparked as a boy. Although Roosevelt suffered from asthma, his home-schooling and love of the outdoors led him to build a collection of natural history specimens obtained in New York and on trips around the world with his family. In his twenties he donated his boyhood collection to the Smithsonian. Later he traveled west to hunt, and after his wife and mother both died on the same day in 1884, he bought a working ranch in North Dakota. He lived there for two years before his hard work was thwarted by snowstorms that decimated his cattle herds. Such down-to-earth experiences taught him to love the land and to appreciate how delicate nature's balances are.

Successive generations have been the beneficiaries of Roosevelt's efforts to save remnants of the great American western wilderness by creating the national park and monument systems. I was proud to serve as secretary of the Smithsonian in 2012 when the Smithsonian and the National Park Service signed an agreement for the Smithsonian to take over the care and preservation of thousands of the artifacts that have been found at our national parks and national monuments.

But the discussion about preserving natural land did not end with Teddy Roosevelt. By the time Jimmy Carter became president, it was focused on the question of how much land should be set aside for protection as wildlife areas and national parks in Alaska, our nation's "last frontier." Congress was generally unified in believing that additional land should be preserved, but Alaska's powerful senators were opposed to excluding large tracts from development.

Like Roosevelt, Carter had learned his love for the land and its flora and fauna by living with it as a child. Like Roosevelt, his appreciation for nature was heightened by hunting and fishing. Like Roosevelt, he set out as president to protect America's wilderness for posterity. And like Roosevelt, his ideas were opposed by many powerful interests when they were first introduced. Despite that opposition, in 1978 Carter signed the National Parks and Recreation Act, which expanded the national park system by almost two million acres by adding fifteen new parks, historic sites, and recreation areas. The legislation also included the Redwood Act, which increased protection for the California parks that contained old-growth redwood groves.

Although the 1978 act was a major step, it left the issue of protected lands in Alaska unsettled. Feeling that time was of the essence in guarding more land against development, Carter and his secretary of the interior, Cecil Andrus, used the provisions of Roosevelt's Antiquities Act to unilaterally set aside more than fifty million acres of land in Alaska for future parks and wildlife areas. The matter was finally resolved when Carter signed the Alaska National Interest Lands Conservation Act into law in 1980 as his term was waning. In one fell swoop, more than a hundred million acres of land were protected, ten new national parks were created, and ecologically sensitive rivers and estuaries were shielded from development. Not everyone was happy. Carter was vilified by hunters, miners, developers, and politicians. He was even criticized by some environmentalists who felt the law did not go far enough. But he stuck by his principles and his love of the land. As Carter said in his foreword to the beautiful photographic book by Subhankar Banerjee on the Arctic National Wildlife Refuge, "There are few places on earth as wild and free as the Arctic Refuge. It is a symbol of our natural heritage, a remnant of frontier America that our first settlers once called wilderness. Little of that precious wilderness remains."

Today, the Smithsonian collections contain more than a thousand books, documents, images, and specimens related to the national parks of Alaska, of which ten owe their existence to a man from South Georgia. I have visited many of the parks and wildlife areas protected when Carter was president. The best times were with my children and grandchildren, knowing that they could still see wilderness and that it will be there for their children. Millions of Americans share this view, and they flock to see these wild places. As Jimmy Carter said when he signed the Alaska Lands Act, "With this bill we are acknowledging that Alaska's wilderness areas are truly this country's crown jewels."

## Ending on a High Note

A good birder can identify a bird as easily by the notes of its song as by sight. Jimmy Carter is such a person. When the Smithsonian Migratory Bird Center and the 2016 North American Ornithological Conference joined forces to announce the publication of the 2016 *State of North America's Birds* report, Carter wrote to support the effort. Titled "Why Birds Really Matter," his blog post stated, "As an avid birder, I know how rewarding it is to watch birds, but it is crucial we also watch over them." He stated that more than one-third of North American birds are at risk of extinction if we do not undertake appropriate conservation steps. Even in his nineties, Carter has continued to work to save the natural world he experienced as a child in South Georgia.

In July 2017 I was fortunate to interview President Carter at the Carter Center in Atlanta, shortly after he returned from a visit to Canada, where he was helping build homes for Habitat for Humanity. He greeted me with a strong handshake and a Carter smile when I entered his office. Our conversation was about the future, and the discussion always came back to South Georgia. We spoke about the beauty of the forests that naturalist William Bartram saw in South Georgia in 1777 and what might be done to preserve some of what is left. He was encouraged by conservation efforts by private organizations and private landowners working in concert with the federal government through its Longleaf Pine Initiative to restore longleaf pine forests. He had recently planted several hundred acres of his own land in longleaf pine. During his lifetime he has seen a comeback of deer and turkey, two species decimated by overhunting and habitat loss, because of focused conservation efforts. He noted with pride that his father, Earl, had been an early advocate for the soil conservation policies promoted during Franklin Roosevelt's administration and that the terraces used in his fields to prevent soil erosion can still be seen today. He hoped we would continue to remember the writings of Bartram to inspire us to do more to save our natural heritage.

When I mentioned his new solar field, his eyes lit up, and he said that although market forces would drive the pace of development, South Georgia was a perfect place for solar energy to thrive and benefit its citizens with clean power. He noted with some irony that he had set aside fifty acres of his land for his solar array, but was limited by the Georgia Power Company grid requirements to using only thirteen.

President Jimmy Carter with the author at the Carter Center, Atlanta, Georgia, July 18, 2017.

The former president spoke of the value of a heritage of the right kind. He is proud of his collection of Native American artifacts, now held at the museum of the Jimmy Carter Presidential Library, and noted that his father and his children had also had a hand in the collecting. He pointed out that all of the artifacts came from his family's farm; he had accepted none from outside collectors. I could sense he felt a connection through his collection to the native people who had lived on the land for thousands of years before European settlers arrived.

Carter hoped that his hometown's history could be maintained through the sites created by the National Park Service, including his boyhood farm and key locales in downtown Plains. He cited the people of Plains as one of the reasons the town has survived while other small rural communities in South Georgia have declined. He believed that the strength of Plains derived from the common bonds of the people who live there and from their history of positive race relations. There was no doubt in his mind that change was needed to make progress, but he thought that Plains had wisely heeded the advice of his teacher and high school principal, Julia Coleman, who said, "We must adjust to changing times, but still hold to unchanging principles."

When I asked him what single factor was the most important in pulling the rural South out of its dark days of poverty and lack of opportunity, he replied without hesitation: "Civil rights and equality for all races." He commented that President Harry Truman's decision in July 1948 to issue an executive order to desegregate the armed forces had lit the fuse, but the back of resistance was broken in the 1950s by the courage of Dr. Martin Luther King Jr. and his nonviolent movement.

My interview with Jimmy Carter left me with a lot to think about. I had spoken with a fellow South Georgian who has lived a life filled with events and experiences unimagined by most of us. Yet all he did and said reflected his roots in the soil where he was born. I was reminded of the closing lines of his book *A Full Life*, subtitled *Reflections at Ninety*: "Times change, and deeds to the various fields will be held in the future by one family and then another. They may or may not be my descendants, but I am confident that the earth itself will remain basically the same, continuing to shape the lives of its owners, for good or ill, as it has for millennia. After all, the land belonged to the Indians before it was ours."

## CHAPTER FIFTEEN

# Reflections

We carry childhood memories with us throughout our lives. When you get to be as old as I am, they come from another time—old black-and-white photographs, ancient-looking cars, and people now gone. When I was growing up in Douglas, Georgia, in the 1940s and early 1950s, it was the end of an era. In the decade to come, segregated facilities would be outlawed, rural homes would be electrified, streets would be paved, kids would no longer be tested for hookworms, and a television would be found in most homes. These changes were needed or inevitable, but I found myself nostalgic for the times when summer days were spent fishing and roaming the woods, when delicious boiled peanuts came steaming off Mom's stove, and when kids knew that disappointing their parents was more damning than anything the authorities might do.

After moving away at the age of twelve, I never lived in South Georgia again, but the dust of the place never left my feet. It stuck with me as I traveled the world to study earthquakes, designed metros in major cities, became a university president, and served as secretary of the Smithsonian. There were stories about growing up in South Georgia I wanted to tell, but it was not virgin territory. Other writers had already been there. Finally it dawned on me. I was secretary of the Smithsonian, an institution whose collections hold millions of artifacts, objects, and specimens. Why not explore them to see what insights would emerge? The challenge was that the massive Smithsonian collections in all their glory loomed large before me. The experts who used them knew what they were looking for, but the collections had never been explored by an amateur who hoped to understand a place called South Georgia. Sometimes it helps to begin a quest without knowing how challenging it is going to be. In my case, that was a blessing in disguise.

My hunting grounds included every one of the twenty units that hold and manage collections at the Smithsonian, and there were more than a few surprises. First, I somehow thought that finding an object from South Georgia in the collections would be the end of the story. Not a chance, because the object itself either blew up some preconceived notion or raised as many questions as it answered.

Second, I did not anticipate the importance of the people in the equation. Every object had been collected by someone, and that someone was often an unusual person. Even after the object finally took its place at the Smithsonian, interesting people called curators studied and took care of it. The people often became as much the story as the object that was collected.

Third, going in I assumed that the documentation about the artifacts in the collections was infallible, but it turns out that even the Smithsonian is human, especially when it comes to the early collections. The institution had few curators in its formative days, and objects flowed over the transom from donors and volunteers. Documentation was sometimes scanty and could not always be vetted. Errors slipped in, some even becoming myth over time. At times the search to sort it all out was like a detective story with clues uncovered along the way.

In another twist, the collections were occasionally found wanting; they were missing something I expected to find there. Twice, my childhood friend Frankie Snow obliged me by providing specimens to fill gaps. When I learned that the National Rock and Ore Collections had no Altamaha grit, which is as important to South Georgia as granite is to Vermont, he supplied not just one specimen but three. When I discovered that the Smithsonian Gardens had no green-fly orchid among its thousands of live orchids, Frankie sent one from Broxton Rocks. When I found the Smithsonian Libraries to be lacking copies of John Muir's *A Thousand-Mile Walk to the Gulf* and Caroline Miller's Pulitzer Prize–winning *Lamb in His Bosom* about the struggles of South Georgia's early settlers, I made up the deficit by gifting copies of early editions to the libraries.

Despite such small errors and omissions, my admiration for the collections grew as each artifact turned out to be a lens to see a new perspective. At every turn the collections turned what seemed to be a "fact" into a story. I learned not only what is known, but also what is not known. I discovered that I was not seeking just to know more about a particular place at a particular time, but to learn how the place had become what it was and how its heritage might be saved for future generations.

Perhaps even more important than my own experience with the Smithsonian collections is the realization that anyone can expand their horizons by doing what I did. Of course, I had special personal access to the collections and to curators willing to help. But with the rapid growth in the institution's digital collections, the day will soon arrive when anyone who is interested in learning or exploring will have online access to the full breadth and depth of the collections.

What did I learn about my home in South Georgia? The collections demonstrate that the South Georgia I knew as a child was the result of large and small forces. The large influences included the rising and falling of sea levels over millions of years, the laying down of sand and sediment to create the coastal plain, the land of South Georgia. Large also was the joining of the continents of North and South America at the Isthmus of Panama some three million years ago, allowing some of the animals we see in South Georgia today, such as the opossum and the armadillo, to trek up from South America and make new homes in the region. The small forces went unnoticed in their time, but their cumulative effects often turned out to be catastrophic. The great megafauna that existed in Georgia fifteen thousand years ago were killed one by one by early humans, but over time their numbers dwindled into extinction.

The Mississippian people who came east to Georgia near the end of the first millennium CE brought cultivated agriculture and large-scale mound building, and they began the process of clearing forests and using regulated burning. But nothing could have prepared them for the arrival of Europeans. New strains of bacteria against which the indigenous people had no defenses caused disease, and the great Native American cultures began to collapse and disappear. From the remnants came constructed communities, such as the Cherokees and Creeks (or Muscogees). Settlers, primarily from Great Britain and the northern American colonies, flooded onto the land. They greatly escalated the process of cutting the forests for timber and clearing the land for crops. The invention of the cotton gin stimulated another round of opportunity seekers, who used enslaved people from Africa to do the hard labor in the fields. The South Georgia that Bartram and Catesby saw in the 1700s disappeared. The Carolina parakeet, the passenger pigeon, and the ivory-billed woodpecker were driven to extinction.

Wars that occurred as the new nation was being formed took their toll even as they determined the ownership of the land. In less than a century, five wars—the Revolutionary War, the War of 1812, and the three Seminole Wars—shifted the land of South Georgia from the hands of Native Americans to those of European descent, who created the United States of America. It was in this time that my family's destiny connected to South Georgia when my great-great-grandfather migrated from New Hampshire to live in Waycross. In 1861 Georgia seceded to join the Confederate States of America, and it took the Civil War to bring it back into the Union in 1865. Members of my family fought on both sides of the war, and many died. The Civil War lasted only four years, but its effects left South Georgia destitute for years to come. Both poor whites and newly freed black people were dependent on subsistence agriculture and lacked educational opportunity. These conditions spawned reactionary laws that legalized segregation and thwarted efforts to revive the South and its economy.

Late in the 1800s Henry Grady's vision for the New South, built on an economy with an industrial base and investment in education, brought optimism to a region that had not felt it for many years. Although segregation undercut its potential, the New South helped to spawn a cultural renaissance that succeeded in spite of the odds. As the twentieth century approached, artists from South Georgia emerged to challenge the established order. Collectively, writers, artists, and musicians exposed the racism that undergirded segregation and demonstrated that whites and blacks shared a common heritage. While it took years and sacrifices by many to bring about the changes needed to create a society based on equal opportunity for all, it did eventually arrive.

The story of Jimmy Carter, who became president of the United States and won the Nobel Peace Prize, suggests that anything is possible for a young person born in South Georgia. Encouraging efforts to save ancient species from extinction, such as bobwhite quail, bears, wild turkeys, alligators, indigo snakes, and longleaf pine, show what is possible if people of goodwill work together. Where old crops, such as tobacco, have faded away, new ones, such as blueberries and olives, are succeeding. Clean solar power, enabled by an abundant year-round supply of sunshine, is competing with coal-fired power plants. Finally, more and more towns and communities are celebrating their patrimony by building museums and heritage centers. Although optimism for the future of South Georgia is encouraging, much remains to be done. The legacy of poverty, lack of opportunity, and ignorance, which darkened the land for so long, will not be overcome in a few years or even generations.

Thomas Wolfe told us that we can't go home again, and he was right in the sense that both Douglas and I changed in the decades since I left. But writing this book allowed me to return to my home and learn not just about the flora, fauna, and history of the place, but about myself and my family. That learning was never in a straight line and came, often as not, as a surprise. I have been left to ponder how seeming coincidences connect events and people across time and space in ways unexpected. In the words of John Muir, "Pick out anything . . . [and] we find it hitched to everything else."

# Acknowledgments

This book, which attempts to cover science, art, history, and culture, could not have happened without support from a wide array of friends and scholars who were willing to suspend their disbelief when they heard about the project. Although I have made every effort to get things right and had much assistance in reviews for this purpose, I take responsibility for any errors that managed to find their way into the final copy.

I begin by offering thanks to my Smithsonian colleagues who patiently helped me appreciate their collections and provided detailed insights into their history. Many also read the chapters in my book related to their collections to help ensure accuracy. In alphabetical order, thanks go to Camille Akeju (deceased), Rachael Allen, David Allison, Alcione Amos, Ben Andrews, Erin Beasley, Bill Billeck, David Bohaska, Sean Brady, Betsy Broun, Cari Corrigan, Torsten Dikow, Carla Dove, Rex Ellis, Marty Emery, Barbara Faust, Valerie Fletcher, Brandon Fortune, Leslie Hale, Chris Helgen, Eric Hollinger, Brian Huber, Jeremy Jacobs, Portia James (deceased), Kirk Johnson, Jennifer Jones, Dennis Kelly, Liza Kirwan, John Kress, Dorothy Lippert, Paul McCutcheon, Ann McMullen, Virginia Mecklenberg, Chris Milensky, David Miller, Scott Miller, Jennifer Morris, Tim Nolan, Silvia Orli, Leslie Overstreet, Daniel Piazza, Jeff Place, Wendy Wick Reaves, Douglas Remley, Torben Rick, Harry Rubenstein, Erin Rushing, Rusty Russell, Jeff Schneider, Floyd Shockley, Ann Shumard, Jane Sledge, Stephanie Smith, Richard Sorenson, Dennis Stanford, Jeffrey Stine, Mary Augusta Thomas, Leslie Umberger, Dennis Whigham, and Jeff Williams.

I owe a debt to other Smithsonian colleagues, including Janet Abrams, who kept score on all of the visits, helped connect me to curators and collections, and never gave up on the project; Johnny Gibbons, who served as the official photographer both in the collections and on field trips; Al

Horvath for being the money man when needed; Patty Bartlett for her encouragement; and Richard Kurin for his wise counsel. Special gratitude goes to the amazing Pam Henson of the Smithsonian Archives for her assistance in finding obscure materials and correspondence and helping me understand the history of the collections. Finally, although she is no longer with us, I would like to thank my friend and former Smithsonian colleague Evelyn Lieberman for her advice and steadfast belief in the project.

Pam Henson comes in for another thank you for introducing me to Lester Stephens, professor emeritus of history at the University of Georgia. Lester provided valuable knowledge about the history of science in the coastal areas of the Southeast and insights into the remarkable LeConte family. He was kind enough to review the book as well. Also deserving thanks is my friend Chuck Knapp, president emeritus of the University of Georgia, who encouraged me to stay the course in the face of challenges.

I offer my appreciation to the Board of Regents of the Smithsonian for their endorsement of the project and providing financial support for it. My gratitude also to the University of Georgia Press for being willing to publish the work of a former president of Georgia Tech. Thanks especially to Thomas Roche, Erin New, and Steven Wallace of the press for editorial and design support.

At my Atlanta base I owe a debt to Ellen Cormack, my first assistant at Georgia Tech after my retirement from the Smithsonian, for her steadfast support and help. Carol Hogan stepped in admirably when Ellen retired and I am grateful to her. From the beginning, Sarah Eby-Ebersole served as my stalwart editor, editing the book more than once and improving its readability, composition, and grammar. I cannot thank her enough.

The chapter on President Jimmy Carter benefited from the assistance of Phil Wise and Stephen Hochman of the Carter Center and of David Stanhope, Sylvia Naguib, and Carla Ledgerwood of the Jimmy Carter Presidential Library and Museum. Of course, thanks go to President Carter, who was not only kind enough to provide time for my interview, but helped me gain a new appreciation for my South Georgia home.

To my childhood friend Frankie Snow, my thanks for all of your advice, insights, willingness to use your connections, and graciousness in joining me on trips to small places in South Georgia. I am lucky to know you. It was through Frankie that I was able to meet Janisse Ray, an author whose works about South Georgia helped me appreciate how much I had to learn about my home.

And to my wife, Anne Robinson Clough, who put up with all of my complaints, trips, extended absences from family activities, and interminable stories about my findings, I cannot say enough about your help, encouragement, and love.

Finally, I am grateful for having been born to loving parents who made sacrifices in their own lives to promote mine. Growing up in South Georgia provided me with a home that nurtured me for what was to come in my life.

# Bibliography

The following are the sources I consulted for each chapter. In addition, the *New Georgia Encyclopedia* was a valuable resource, which I recommend to interested readers who would like to learn more about the people, places, and events I discuss. See http://www.georgiaencyclopedia.org. The website of the Smithsonian Collections Search Center, http://collections.si.edu/search, provides more detailed information about the collections.

### *Introduction*

Muir, John. *My First Summer in the Sierra*. Boston: Houghton Mifflin, 1911.

### *Chapter 1. Recollections of a South Georgia Childhood*

Georgia Forestry Commission. "Tree Talk: Village Sentinel." http://www.gatrees.org/TreeTalk/Archive/VillageSentinel-Waycross.cfm.

Griffiths, Leann Elizabeth. *Images of America: Douglas*. Charleston, S.C.: Arcadia, 2014.

Ward, Warren P. *Ward's History of Coffee County*. Atlanta: Foote and Davies, 1930.

Wong, Eva, trans. *Lieh-tzu: A Taoist Guide to Practical Living*. Boston: Shambhala Publications, 2001.

### *Chapter 2. The Process of Discovery*

Henson, Pamela. "A National Science and a National Museum." In *Museums and Other Institutions of Natural History: Past, Present and Future*, ed. A. E. Leviton and M. L. Aldrich, 34–57. San Francisco: California Academy of Sciences, 2004.

Morton, William J. *The Story of Georgia's Boundaries: A Meeting of History and Geography*. Atlanta: Georgia History Press, 2009.

Philbrick, Nathaniel. *The Scientific Legacy of the U.S. Exploring Expedition*. Washington, D.C.: Smithsonian Institution Libraries, January 2004.

Ricard, Serge, ed. *A Companion to Theodore Roosevelt*. Hoboken, N.J.: Wiley, 2011.

*Chapter 3. Beginning with Fossils*

Bedini, Silvio A. *Thomas Jefferson and American Vertebrate Paleontology*. Charlottesville: Virginia Division of Mineral Resources, 1985.

Foreman, E. "Notice of Screven Fossil Donation to National Institute." *National Intelligencer*, September 9, 1842.

Harden, William. *A History of Savannah and South Georgia*. 2 vols. Chicago: Lewis Publishing, 1913.

Hodgson, William B. *Memoir on the Megatherium and Other Extinct Gigantic Quadrupeds of the Coast of Georgia with Observations on Its Geologic Features*. New York: Bartlett and Welford, 1846.

Leidy, Joseph. "A Memoir on the Extinct Sloth Tribe of North America." *Smithsonian Contributions to Knowledge* (December 1853): 49–55.

"*Megalonyx jeffersonii* Fossils." In *Thomas Jefferson Encyclopedia*. https://www.monticello.org/site/research-and-collections/megalonyx-jeffersonii-fossils.

Naish, Darren. "The Anatomy of Sloths." *Scientific American*, August 30, 2012. https://blogs.scientificamerican.com/tetrapod-zoology/the-anatomy-of-sloths.

Patterson, David B., Alfred J. Mead, and Robert Bahn. "New Skeletal Remains of *Mammuthus columbi*, from Glynn County, Georgia with Notes on Their Historical and Paleoecological Significance." *Southeast Naturalist* 2, no. 2 (2012): 163–72.

Rudwick, Martin J. S. *Georges Cuvier, Fossil Bones, and Geological Catastrophes*. Chicago: University of Chicago Press, 2009.

Stephens, Lester D. *Science, Race, and Religion in the American South: John Bachman and the Charleston Circle of Naturalists, 1815–1895*. Chapel Hill: University of North Carolina Press, 2013.

*Chapter 4. Book Learning*

Alderfer, Jonathan. "Celebrating the Remarkable Mark Catesby." *National Geographic*, September 25, 2012. http://voices.nationalgeographic.com/2012/09/25/celebrating-the-remarkable-mark-catesby.

Bartram, William. *Travels through North and South Carolina, Georgia, East and West Florida, the Cherokee Country, the Extensive Territories of the Muscogulges or Creek Confederacy, and the Country of the Choctaws*. Philadelphia: James and Johnson, 1791.

Catesby, Mark. *The Natural History of Carolina, Florida, and the Bahama Islands*. 2 vols. London, 1731, 1743.

Muir, John. *A Thousand-Mile Walk to the Gulf*. Edited by William Frederic Badè. Boston: Houghton Mifflin, 1916.

Smith, James Edward. *The Natural History of the Rarer Lepidopterous Insects of Georgia . . . Collected from the Observations of Mr. John Abbot, Many Years a Resident of That Country*. London: T. Bensley, 1797.

*Chapter 5. Mammals*

"Continental Drift." University of California, Berkeley. http://www.ucmp.berkeley.edu/geology/anim1.html.

Flannery, Tim. *The Eternal Frontier: An Ecological History of North America and Its Peoples*. New York: Atlantic Monthly Press, 2001.

Tekiela, Stan. *Mammals of Georgia Field Guide*. New York: Advance Publishing, 2011.

*Chapter 6. Birds and the Stories They Tell of the People Who Collect Them*

Burleigh, Thomas D. *Georgia Birds*. Norman: University of Oklahoma Press, 1958.

Catesby, Mark. *The Natural History of Carolina, Florida, and the Bahama Islands*. 2 vols. London, 1731, 1743.

"Dr. Joseph LeConte." Sierra Club. https://www.sierraclub.org/yosemite-heritage-center/dr-joseph-leconte.

Gallagher, Tim. *The Grail Bird: Hot on the Trail of the Ivory-Billed Woodpecker*. Boston: Houghton Mifflin, 2005.

"The Impact of Joseph LeConte (1869–1901)." University of California Museum of Paleontology. November 2006. http://www.ucmp.berkeley.edu/about/history/lipps2.php.

LeConte, Joseph. *The Autobiography of Joseph LeConte*. Edited by William Dallas Ames. New York: Appleton, 1903.

———. "Memoir of John LeConte, 1818–1891." Read before the National Academy of Sciences, Washington, D.C., April 1894. http://www.nasonline.org/publications/biographical-memoirs/memoir-pdfs/le-conte-john.pdf.

Stephens, Lester D. *Joseph LeConte: Gentle Prophet of Evolution*. Baton Rouge: Louisiana State University Press, 1982.

———. "Joseph LeConte's Contribution to American Ornithology." *Georgia Journal of Science* 35 (June 1977): 170–81.

Stoddard, Herbert L. *The Bobwhite Quail: Its Habits, Preservation, and Increase*. New York: Scribner, 1931.

Weatherford, Carole Boston. *The Carolina Parakeet: America's Lost Parrot in Art and Memory*. Fridley, Minn.: Avian Publications, 2005.

White, George. *Statistics of the State of Georgia, Including an Account of Its Natural, Civil and Ecclesiastical History*. Savannah: W. Thomas Williams, 1849.

Yeoman, Barry. "Why the Passenger Pigeon Went Extinct." *Audubon*, May–June 2014.

*Chapter 7. Amphibians, Reptiles, and a Detour into My Family History*

Anderson, Mary Ann, ed. *Cen-Silver Celebration: The 125th Anniversary of the Founding of Hazelhurst, Georgia, a Town at the Crossroads*. Hazelhurst, Ga.: Hazelhurst–Jeff Davis County Chamber of Commerce, 1995.

"Broxton Rocks Preserve." Nature Conservancy. http://www.nature.org/ourinitiatives/regions/northamerica/unitedstates/georgia/placesweprotect/broxton-rocks.xml.

Jensen, John B., et al. *Amphibians and Reptiles of Georgia*. Athens: University of Georgia Press, 2008.

Kollipara, Puneet. "Climate Change Shrinks Salamander." *National Geographic*, March 27, 2014. http://news.nationalgeographic.com/news/2014/03/140327-climate-change-shrinks-salamanders-global-warming-science.

Ruckdeschel, C., C. R. Shoop, and G. R. Zug. *Sea Turtles of the Georgia Coast*. St. Marys, Ga.: Cumberland Island Museum, 2000.

U.S. Fish and Wildlife Service. "Conserving Amphibians: What the Amphibians Are Telling Us and Why We Should Listen." https://www.fws.gov/Endangered/news/amphibians.html.

*Chapter 8. So Many Specimens, So Little Time*

Bishop, G. Norman. *Native Trees of Georgia*. Rev. ed. Georgia Forestry Commission, 2001.

Dillard, Annie. *Pilgrim at Tinker Creek*. New York: HarperCollins, 1999.

Grissell, E. E., and Thomas R. Fasulo. "Yellowjackets and Hornets." University of Florida Featured Creatures Collection. http://edis.ifas.ufl.edu/in238.

Hobbs, Horton H., Jr. *The Crayfishes of Georgia*. Washington, D.C.: Smithsonian Institution Press, 1981.

Muir, John. *A Thousand-Mile Walk to the Gulf*. Edited by William Frederic Badè. Boston: Houghton Mifflin, 1916.

Shores, Elizabeth Findley. *On Harper's Trail: Roland McMillan Harper, Pioneering Botanist of the Southern Coastal Plains*. Athens: University of Georgia Press, 2010.

*Chapter 9. Geology*

American Meteor Society. "Fireball FAQs." https://www.amsmeteors.org/fireballs/faqf.

Flannery, Tim. *The Eternal Frontier: An Ecological History of North America and Its Peoples*. New York: Atlantic Monthly Press, 2001.

Ford, Latasha. "Men Meet to Discuss Enigma Meteorite." *Tifton Gazette*, April 26, 2015. http://www.tiftongazette.com/news/men-meet-to-discuss-enigma-meteorite/article_16db2106-eb7e-11e4-a062-0b6cdd2a65d5.html.

"Georgia Meteorites." Meteorite Association of Georgia. http://www.meteoriteassociationofgeorgia.org/GAMeteorites.htm.

Huddlestun, Paul F. "A Revision of the Lithostratigraphic Units of the Coastal Plain of Georgia: The Miocene through Holocene." *Georgia Geological Survey Bulletin* 104 (1988): 100–112.

Krakow, Kenneth K. *Georgia Place Names: Their History and Origins*. Macon, Ga.: Winship Press, 1975.

McCallie, S. W. "The Pitts Meteorite." *American Journal of Science and Arts*, 5th ser., 3 (1922): 211–15.

Norton, O. Richard. *Rocks from Outer Space: Meteorites and Meteorite Hunters*. 2nd ed. Missoula, Mont.: Mountain Press, 2001.

Osier, David R. "A Reprieve for the Okefenokee Swamp?" *Southerner* 1, no. 4 (Fall 1999). http://www.southerner.net/v1n4_99/cover4.html.

Plotkin, H., and R. S. Clark Jr. "Stuart H. Perry's Contributions to Meteorite Collection and Research, 1927–1957." *Meteorites and Planetary Science* 44, no. 8 (2009): 1161–77.

*Chapter 10. Early Native American Civilizations*

Bartram, William. *Travels through North and South Carolina, Georgia, East and West Florida, the Cherokee Country, the Extensive Territories of the Muscogulges or Creek Confederacy, and the Country of the Choctaws*. Philadelphia: James and Johnson, 1791.

Blanton, Dennis, et al. *Points of Contact: The Archaeological Landscape of Hernando de Soto in Georgia*. Atlanta, Ga.: Fernbank Museum of Natural History, 2011.

Clayton, Lawrence A., James Knight Jr., and Edward C. Moore, eds. *The De Soto Chronicles*. 2 vols. Tuscaloosa: University of Alabama Press, 1995.

Elliott, Daniel T. *Purchased, Given, Exchanged, Condemned, Destroyed and Neglected: Discovery of the McGlashan Collection at the Smithsonian Institution*. Savannah, Ga.: LAMAR Institute, 2010.

———. *The Roland Steiner Project, Phase I: Introductory Background and Correspondence*. Savannah, Ga.: LAMAR Institute, 2010.

Hally, David J., ed. *Ocmulgee Archaeology, 1936–1986*. Athens: University of Georgia Press, 1994.

Jones, Charles Colcock, Jr. *Antiquities of the Southern Indians, Particularly of the Georgia Tribes*. New York: Appleton, 1873.

Kelly, A. R., and Betty A. Smith. *The Swift Creek Site, 9Bi3*. Macon, Ga.: Ocmulgee National Monument, 1975.

King, Adam. "The Lord of Death on the Savannah River." *University of South Carolina Anthropology Commons* 17, no. 1 (2013).

Mann, Charles C. *1491: New Revelations of the Americas before Columbus*. New York: Knopf, 2005.

Snow, F. H. "Swift Creek Designs and Distribution: A South Georgia Study." *Early Georgia* 3, no. 2 (December 1975): 38–53.

Thomas, Cyrus. *Report on the Mound Explorations of the Bureau of Ethnology*. Washington, D.C.: Smithsonian Institution Press, 1894.

Williams, Mark, and Daniel T. Elliott, eds. *A World Engraved: Archaeology of the Swift Creek Culture*. Tuscaloosa: University of Alabama Press, 1998.

Williams, Mark, and Keith Stephenson. "Henry L. Reynolds, Jr.: Georgia's Greatest 19th Century Archaeologist." *Early Georgia* 40, no. 1 (Spring 2012): 69–86.

*Chapter 11. The End of One Era Leads to the Beginning of Another*

Faust, Drew Gilpin. *This Republic of Suffering: Death and the American Civil War.* New York: Knopf, 2008.

Fordney, Ben Fuller. *George Stoneman: A Biography of the Union General.* Jefferson, N.C.: McFarland, 2008.

Grant, Ulysses S. *Personal Memoirs of U. S. Grant.* New York: Charles E. Webster, 1885.

Green, Constance M. *Eli Whitney and the Birth of American Technology.* Boston: Little, Brown, 1956.

Groce, W. Todd. "Rethinking Sherman's March." *New York Times,* November 17, 2014, https://opinionator.blogs.nytimes.com/2014/11/17/rethinking-shermans-march.

Kantor, Mackinlay. *Andersonville.* New York: Penguin, 1955.

Levy, George. *To Die in Chicago: Confederate Prisoners at Camp Douglas, 1862–1865.* Louisville, Ky.: Evanston, 1994.

Lewis, Lloyd. *Sherman: Fighting Prophet.* Lincoln: University of Nebraska Press, 1993.

Monroe, Kristopher. "The Weeping Time: A Forgotten History of the Largest Slave Auction Ever on American Soil." *Atlantic,* July 10, 2014. http://www.theatlantic.com/business/archive/2014/07/the-weeping-time/374159.

Morton, William J. *The Story of Georgia's Boundaries: A Story of History and Geography.* Atlanta: Georgia History Press, 2009.

Nelson, Megan Kate. *Trembling Earth: A Cultural History of the Okefenokee Swamp,* Athens: University of Georgia Press, 2005.

Pierce, John. "The Reasons for Secession: A Documentary Study." American Battlefield Trust. http://www.civilwar.org/education/history/secession.

Ransom, John L. *John Ransom's Andersonville Diary.* Philadelphia: Douglass Brothers, 1883.

Remini, Robert V. *Andrew Jackson and His Indian Wars.* New York: Viking Penguin, 2001.

Rigdon, John C. *Historical Sketch and Roster of the Georgia 26th Infantry Regiment.* Cartersville, Ga.: Eastern Digital Resources, 2013.

Speare, Eva Clough, ed. *The Genealogy of the Descendants of John Clough of Salisbury, Massachusetts.* Exeter, N.H.: John Clough Genealogical Society, 1952.

Stewart, Richard W., general ed. *American Military History,* vol. 1: *The United States Army and the Forging of the Nation.* Washington D.C.: Center of Military History, U.S. Army, 2005.

*Chapter 12. Gullah Geechee*

Conroy, Pat. *The Water Is Wide.* New York: Random House, 1987.

Cross, Wilbur. *Gullah Culture in America.* Winston-Salem, N.C.: John E. Blair Publisher, 2012.

*Word, Shout, Song: Lorenzo Dow Turner Connecting Communities through Language.* Exhibition catalog, August 9, 2010–July 24, 2011.

*Chapter 13. South Georgia Art and Artists*

Davis, Harold E. *Henry Grady's New South: Atlanta, a Brave and Beautiful City.* Tuscaloosa: University of Alabama Press, 1990.

Goode, G. Brown. "Report upon the Exhibit of the Smithsonian Institution and the United States National Museum at the Cotton States and International Exposition, Atlanta, Ga., 1895." In *Smithsonian Report for 1896*, 613–35 and plate LXI. Washington, D.C.: U.S. Government Printing Office, 1898.

Jean-Laurent, Annabella. "Flashback: The 1895 Cotton States Exposition and the Negro Building." *Atlanta*, February 27, 2014. http://atlantamagazine.com/news-culture-articles/flashback-the-1895-cotton-states-exposition-and-the-negro-building.

Kennedy, Roger, and Ann Prentice Wagner. *1934: A New Deal for Artists.* London: Giles, 2009.

Mims, Edwin. *A Biography of Sidney Lanier.* Boston: Houghton Mifflin, 1905.

Rosenberg, Karen. "A Barber's Carved Legacy, Finished with Rhinestones and Shoe Polish." *New York Times*, April 23, 2009. http://www.nytimes.com/2009/04/24/arts/design/24davi.html.

*Chapter 14. Jimmy Carter*

Bourne, Peter G. *Jimmy Carter: A Comprehensive Biography from Plains to Post-Presidency.* New York: Scribner, 1997.

Carter, Jimmy. Foreword. In Subhankar Banerjee, *Arctic National Wildlife Refuge: Seasons of Life and Fire.* Seattle, Wash.: Mountaineers Books, 2003.

———. *A Full Life: Reflections at Ninety.* New York: Simon and Schuster, 2015.

———. *A Government as Good as Its People.* 1977. Reprint, Fayetteville: University of Arkansas Press, 1996.

———. *An Hour before Daylight: Memories of a Rural Boyhood.* New York: Simon and Schuster, 2001.

———. *An Outdoor Journal: Adventures and Reflections.* New York: Bantam, 1988.

———. "Why Birds Really Matter: President Jimmy Carter." *Smithsonian Insider*, August 20, 2016. https://insider.si.edu/2016/08/birds-really-matter-4.

Carter, Rosalynn. *First Lady from Plains.* Fayetteville: University of Arkansas Press, 1994.

Hemauer, Christina, and Roman Keller, dirs. *A Road Not Taken: The Story of the Jimmy Carter White House Solar Installation.* Documentary film, 2010.

Stine, Jeffrey K. "Jimmy Carter's Solar Vision." *Stories from the National Museum of American History*, Smithsonian Institution, April 23, 2014. http://americanhistory.si.edu/blog/2014/04/jimmy-carters-solar-vision.html.

Zelizer, Julian E. *Jimmy Carter.* New York: Macmillan, 2010.

# Illustration Credits

Page 9, Clough Ice and Coal plant, photographer unknown, courtesy Ethan S. M. Fellows.

Page 11, downtown Douglas, Coffee County centennial celebration, photograph by Bernard Rogers of Rogers Studio, Douglas, Georgia.

Page 19, "Village Sentinel" live oak at the Baptist Retirement Home, photograph by Johnny Gibbons.

Page 22, South Georgia gnat, *Liohippelates bishoppi*, National Museum of Natural History entomology collections, photograph by Johnny Gibbons.

Page 26, bird collections, National Museum of Natural History, photograph by Chip Clark.

Page 32, *Eremotherium* fossil in the Dinosaur Hall, National Museum of Natural History vertebrate paleontology collections, photograph by Johnny Gibbons.

Page 34, *Eremotherium* fossil bones from Skidaway Island, National Museum of Natural History vertebrate paleontology collections, photograph by Johnny Gibbons.

Page 34, *Eremotherium* fossil jawbone from Skidaway Island, National Museum of Natural History vertebrate paleontology collections, photograph by Johnny Gibbons.

Page 42, "parrot of Carolina" and "cypress of America" from Mark Catesby, *The Natural History of Carolina, Florida and the Bahama Islands*, vol. 1 (1731), Smithsonian Libraries rare book collections.

Page 43, ivory-billed woodpecker and willow oak, from Mark Catesby, *The Natural History of Carolina, Florida and the Bahama Islands*, vol. 1 (1731), Smithsonian Libraries rare book collections.

Page 46, "Mico Chlucco the Long Warrior, or King of the Siminoles," from William Bartram, *Travels* (1791), Smithsonian Libraries rare book collections.

Page 56, golden mice from South Georgia, National Museum of Natural History mammals collections, photograph by Johnny Gibbons.

Page 60, bat from near LeConte Plantation, Riceboro, Georgia, National Museum of Natural History mammals collections, photograph by Johnny Gibbons.

Page 61, fox squirrels from South Georgia, National Museum of Natural History mammals collections, photograph by Johnny Gibbons.

Page 67, brown thrashers from South Georgia, National Museum of Natural History bird collections, photograph by Chris Milensky.

Page 69, ivory-billed woodpecker from Okefenokee Swamp, National Museum of Natural History bird collections, photograph by Chris Milensky.

Page 72, portrait of Joseph LeConte by Louis O. Lussier, 1876, National Portrait Gallery collections, owned by University of California, Berkeley, Bancroft Library.

Page 72, portrait of John LeConte by Louis O. Lussier, 1877, National Portrait Gallery collections, owned by University of California, Berkeley, Bancroft Library.

Page 77, Caroline Herrington and the author at LeConte's Woodmanston Plantation and Botanical Gardens near Midway, Georgia, photograph by Johnny Gibbons.

Page 78, LeConte Norman red camellia at LeConte's Woodmanston Plantation and Botanical Gardens, photograph by Johnny Gibbons.

Page 81, salamanders from South Georgia, National Museum of Natural History amphibian collections, photograph by Johnny Gibbons.

Page 83, eastern diamondback rattlesnake skin from South Georgia, National Museum of Natural History reptile collections, photograph by Johnny Gibbons.

Page 87, Broxton Rocks waterfall, photograph by Usayd Casewit.

Page 89, herpetologist Dirk Stevenson with Frankie Snow and the author, photograph by Johnny Gibbons.

Page 90, the author holding an eastern indigo snake, photograph by Johnny Gibbons.

Page 96, southern yellow jackets from South Georgia, National Museum of Natural History entomology collections, photograph by Johnny Gibbons.

Page 97, cow killers from South Georgia, National Museum of Natural History entomology collections, photograph by Johnny Gibbons.

Page 102, longleaf pine specimen, National Museum of Natural History botany collections.

Page 104, Altamaha grit rock face at Broxton Rocks, photograph by Usayd Casewit.

Page 106, Okefenokee pygmy sunfish, photograph by Blickwinkel/Alamy Stock Photos.

Page 108, Christmas tree crayfish collected in South Georgia and sketched by Horton Hobbs. National Museum of Natural History invertebrate zoology collections.

Page 116, Enigma meteorite, National Museum of Natural History meteorite collections.

Page 119, Pitts meteorite, National Museum of Natural History meteorite collections.

Pages 120–21, downtown Pitts, Georgia, photograph by the author.

Page 123, Nancy Brinson house site, where the largest fragment of the Pitts meteorite fell, photograph by Frankie Snow.

Page 124, Snow Hill Baptist Church, Pitts, Georgia, photograph by Mike Miller.

Page 127, Osco Bennett, photograph by Frankie Snow.

Page 140, earthen bottle, Mississippian, Hollywood site, National Museum of Natural History anthropology collections, photograph by Johnny Gibbons.

Page 140, copper plate, underwater panther, Mississippian, Hollywood site, National Museum of Natural History anthropology collections, photograph by Johnny Gibbons.

Page 143, discoidal (chunkey) stone, Mississippian, National Museum of Natural History anthropology collections, photograph by Johnny Gibbons.

Page 146, Frankie Snow's collection of Swift Creek pottery and paddle images, South Georgia State College, photograph by Johnny Gibbons.

Page 148, Clovis points found in Georgia, National Museum of Natural History anthropology collections.

Page 153, one-cent George Washington and Nathanael Greene small die proof, postage stamp, 1936, National Postal Museum collections.

Page 154, free frank on Valley Forge cover, National Postal Museum collections.

Page 156, reproduction of Eli Whitney's cotton gin, Treasures of American History at the National Museum of American History.

Page 160, William McIntosh (Tustunnuggee Hutkee), lithograph of portrait by Charles Bird King, from Thomas McKenney and James Hall, *History of the Indian Tribes of North America* (1836). National Portrait Gallery collections.

Page 161, portrait of Osceola (William Powell) by George Catlin, 1838, Smithsonian American Art Museum collections.

Page 164, business card of Blount and Dawson, National Museum of African American History and Culture collections.

Page 167, headstone of John D. Clough, photograph by Usayd Casewit.

Page 169, General William T. Sherman, albumen silver print, National Portrait Gallery collections, photographer unknown.

Page 169, pin made of buttons taken from the frock coat of General William T. Sherman, Armed Forces History collections, National Museum of American History.

Page 171, Griswold and Gunnison revolver, National Museum of American History Armed Forces History collections.

Page 173, engagement rings of Varina Howell Davis, wife of Jefferson Davis, National Museum of American History collections.

Page 180, Lorenzo Dow Turner, Anacostia Community Museum collections, photographer unknown.

Page 182, ring shout, Anacostia Community Museum collections, photograph by Lorenzo Dow Turner.

Page 183, aluminum disk used by Lorenzo Dow Turner to record Mende funeral song sung by Amelia Dawley, Anacostia Community Museum collections, photograph by Johnny Gibbons.

Page 190, portrait of Erskine Caldwell by Frederick Wight, 1934, National Portrait Gallery collections, courtesy of Estate of Frederick Wight, reproduced by permission.

Page 190, portrait of Carson McCullers by Irving Penn, 1950, National Portrait Gallery collections, courtesy of Condé Nast, reproduced by permission.

Page 191, portrait of Flannery O'Connor by Joseph Reshower, 1961, National Portrait Gallery collections, courtesy of Joseph Reshower Estate, reproduced by permission.

Page 194, George Inness, *Georgia Pines* (1890), Smithsonian American Art Museum collections.

Page 195, Ulysses Davis, *Beast Going through the Grass* (1984–1985), Smithsonian American Art Museum collections.

Page 197, Alma Thomas, *The Eclipse* (1970), Smithsonian American Art Museum collections.

Page 198, portrait of Alma Thomas by Laura Wheeler Waring, 1945, Smithsonian American Art Museum collections.

Page 201, Daniel Celentano, *The Country Store and Post Office*, study for mural in Vidalia, Georgia, post office, 1938, Smithsonian American Art Museum collections.

Page 203, Gertrude "Ma" Rainey, *Queen of the Blues*, Biograph Records, 1971, Smithsonian Folkways collection.

Page 204, *Blind Willie McTell, 1927–1935*, Yazoo Records, 1991, Smithsonian Folkways collection.

Page 205, *Darby and Tarlton*, Old Timey Records, 1971, Smithsonian Folkways collection.

Page 206, sheet music for "Lazybones," music by Hoagy Carmichael and lyrics by Johnny Mercer, 1932, National Museum of American History.

Page 208, Ray Charles, National Portrait Gallery collections, photograph by Michel Salou.

Page 209, red jumpsuit owned by James Brown, National Museum of African American History and Culture collections, photographer unknown.

Page 210, *Little Richard's Grooviest 17 Original Hits!*, Teichiku Records, 1968, Smithsonian Folkways collections.

Page 217, Carter for President campaign peanut bank, National Museum of American History collections.

Page 218, Environmentalists for Carter presidential campaign pins, National Museum of American History collections.

Page 219, portrait of Jimmy Carter by Alan Reingold, 1976, National Portrait Gallery collections.

Page 222, portrait of Rosalynn Carter by Robert Templeton, 1976, National Portrait Gallery collections.

Page 224, Shanthi (elephant) with Amy Carter, Rosalynn Carter, and Punita Gunaratne, 1977, Smithsonian Institution Archives, no. 77-4014-14.

Page 229, Jimmy Carter with the author, photograph by Phillip Wise.